The Thunderbirds

The United States Air Force's Flight Demonstration Team, 1953 to the Present

KEN NEUBECK

SCHIFFER MILITARY

4880 Lower Valley Road   Atglen, PA 19310

Front cover photo by John Gourley
Rear cover photo by Ken Neubeck
Designed by Justin Watkinson
Type set in Impact/Minion Pro/Univers LT Std

ISBN: 978-0-7643-6076-3
Printed in China

Published by Schiffer Publishing, Ltd.
4880 Lower Valley Road
Atglen, PA 19310
Phone: (610) 593-1777; Fax: (610) 593-2002
E-mail: Info@schifferbooks.com
www.schifferbooks.com

For our complete selection of fine books on this and related subjects, please visit our website at www.schifferbooks.com. You may also write for a free catalog.

Schiffer Publishing's titles are available at special discounts for bulk purchases for sales promotions or premiums. Special editions, including personalized covers, corporate imprints, and excerpts, can be created in large quantities for special needs. For more information, contact the publisher.

We are always looking for people to write books on new and related subjects. If you have an idea for a book, please contact us at proposals@schifferbooks.com.

Acknowledgments

The Thunderbirds have been a regular visitor to the Long Island area to perform at the Jones Beach Air Show. Over the years, I was fortunate to be able to photograph them at Republic Airport and MacArthur Airport on Long Island, where they were based for the air show. The team consisted of pilots and support personnel who always interfaced well with the public. Just like the Blue Angels, who visit the Long Island area for the same air show on alternate years, the Thunderbirds are a major event for the air show as well.

The author thanks those individuals who contributed to the book, including Michael Benolkin, Mike Machat, Josh Stoff, Stu Walerstein, Cradle of Aviation, Lou Drendel, John Gourley, and Dan Wilbur, as well as Audrey Cohen of Epoch.5 Public Relations. The author thanks organizations such as the Thunderbirds Alumni Association and the Long Island Republic Airport Historical Society, and the USAF media websites for their help with historical photographs. In addition, the author thanks the staff at MacArthur Airport on Long Island for the help that they provided to him during the Jones Beach Air Show with regard to access to the Thunderbirds during arrival activities.

Abbreviations used in this book:

CAM = Cradle of Aviation Museum
LIRAHS = Long Island Republic Airport Historical Society
USAF = US Air Force

Contents

Introduction

The Thunderbirds emblem was developed in 1953 at Luke AFB. It originally had the name "Luke" at the top of the emblem. *USAF*

THUNDERBIRDS AIRCRAFT MODELS

Model	Years in Team
Republic F-84G	1953–1954
Republic F-84F	1955–1956
North American F-100C	1956–1963
Republic F-105B	1964
North American F-100D	1964–1968
McDonnell F-4E	1969–1973
Northrop T-38	1974–1982
General Dynamics F-16	1983 to present

The Thunderbirds are officially known as the US Air Force Air Demonstration Squadron. The team's origin was at Luke Air Force Base (AFB) in Arizona, which was the home of the Air Force Advance Flight Training School, and in May 1953 the 3600th Air Demonstration Team was formed. There was no name for the team at the time that it was formed, and a contest for the name for the team was held in 1953, where the name "Thunderbird" was chosen. The name was very much in sync with the Native American folklore in the desert area where the team flew. The Thunderbirds were associated with success in war, and the figure used red, white, and blue colors. From this concept, a formal emblem was developed for the team, along with the label "Luke" on top of the emblem.

The first Thunderbird team consisted of four pilots: Maj. Dick Catledge as the team lead, twins Capt. Bill and Capt. Buck Patillo for the left and right wings, and Capt. Bob Kanaga as slot pilot. The team performed maneuvers with three aircraft for a portion of the show, with the fourth aircraft being used for completing the diamond formation.

Later, the spare aircraft for the team was added as the fifth plane and would be used to perform solo maneuvers that were independent of the four-aircraft formation. The first Thunderbirds solo pilot for the team was Lt. Bob McCormick.

The Thunderbirds' first show was conducted on June 8, 1953, at Luke AFB, and the first civil show was for the Frontier Days event in Cheyenne, Wyoming, in August of that year. The team performed a total of fifty shows during their first year in 1953.

The evolution of Thunderbird aircraft mirrors the advances in aviation over the past sixty-five years—going from propeller-driven aircraft to the F-16 jet engine aircraft currently used by the team.

The Thunderbirds fly throughout the US each year and over a two-year period will cover over sixty different locations in the US and other countries before returning to many of the same locations again. The mission of the squadron is to plan and present precision aerial maneuvers to exhibit the capabilities of modern, high-performance aircraft and the high degree of professional skill required to operate those aircraft.

The first Thunderbird pilots: team leader Maj. Catledge is on the right, along with the two Patillo twins (Captains Bill and Buck) flanking Capt. Kanaga on the left. *USAF*

CHAPTER 1
Republic F-84G Thunderjet

F-84 THUNDERJET SPECIFICATIONS	
Wingspan	36 feet, 5 inches
Length	38 feet, 1 inch
Height	12 feet, 7 inches
Empty weight	11,095 pounds
Power plant	one Allison J35-29 turbojet engine
Maximum speed	622 mph
Service ceiling	41,000 feet
Range	870 miles

The F-84 Thunderjet series was the first production jet fighter aircraft developed by Republic Aviation Corporation for the US Air Force. When the Thunderbirds formed in 1953, they used the last straight-wing model of this series, the F-84G.

The Republic F-84G served during the Korean War and saw a significant amount of activity in attacking ground targets during the war. It was a logical choice for the new Thunderbirds team to use to demonstrate jet aircraft.

The setup for the Thunderbirds was to use three or four aircraft for most of their group maneuvers, and a solo aircraft to fill in between the group maneuvers during the show.

The team was led by Maj. Dick Catledge during the two-year period that this aircraft was used.

There was one accident during the F-84G Thunderbirds period, with Capt. George Kevil being killed during solo training in Arizona in December 1954.

The Thunderbirds flew the F-84G model for two years, from 1953 through 1954, until the F-84F swept-wing version was available and the team switched to that aircraft.

A total of 3,025 F-84G Thunderjets were built for the USAF. The Thunderjet saw significant action during the Korean War and served with the US Air Force through the end of the 1950s. *Republic Archives via LIRAHS*

This is a team photo of the 1953 Thunderbirds in front of the F-84G Thunderjet aircraft used by the team. Located on the lower left are the Thunderbirds team as led by Maj. Catledge and the three members of the diamond formation, along with the spare pilot. They are surrounded by maintenance and other support personnel. *USAF*

This is a photo of the F-84G during a visit to Brazil in 1954. The team would visit several Latin American countries during the F-84G period, and the national flags of these countries would be painted on the side of the fuselage. The individual aircraft in the team did not have numbers painted on the aircraft, in contrast to the numbering scheme of the US Navy Blue Angels. *Brazil National Agency*

Most of the maneuvers and publicity photos for the F-84G Thunderbirds involve four aircraft, and often in the diamond formation as seen here. The only numerical markings that are used to indicate the #1 and the other aircraft are small numbers located on the nose landing-gear door. *USAF*

Another view of the F-84G Thunderbirds' four-aircraft diamond formation. *USAF via LIRAHS*

At the same time that the Thunderbirds were touring the US with the F-84G aircraft, another competing Air Force team was the Skytamers, which were based in USAFE bases in Europe and conducted tours in different countries in Europe. This team was active from 1950 through 1955. *USAF via LIRAHS*

A view of the four-aircraft diamond formation during an air show that featured the F-84G Thunderbirds. *Brazil National Agency*

This is an underside view of the F-84G Thunderbirds in the four-aircraft diamond formation during a practice over the Arizona desert near Luke AFB. A good part of the aircraft was left in the natural-metal finish, with the nose and the tail section being painted in red, white, and blue color scheme, as well as the wing tip fuel tanks. *USAF via LIRAHS*

CHAPTER 2

Republic F-84F Thunderstreak

F-84F SPECIFICATIONS

Wingspan	33 feet, 7.25 inches
Length	43 feet, 4.75 inches
Height	14 feet, 4.75 inches
Empty weight	13,830 pounds
Power plant	one Wright J65-W-3 turbojet engine
Maximum speed	695 mph
Service ceiling	46,000 feet
Range	810 miles

In 1955, the Thunderbirds would proceed to the next stage of aircraft design, with the Republic F-84F Thunderstreak swept-wing jet aircraft being used.

At this time, the team would add an additional solo aircraft to work with the other solo aircraft, to do maneuvers to fill in the time outside the four-aircraft diamond routine.

The F-84F would be the first Thunderbird aircraft that was modified to hold smoke oil tanks, for creating exhaust smoke from the aircraft by injecting a light oil onto the exhaust, for air show effects.

The team would fly the F-84F for only a year and half, from early 1955 to May 19, 1956, for a total of ninety-one shows. At that point, the team went to the supersonic F-100C, making the Thunderbirds the world's first supersonic aerial demonstration team. There were no crashes or fatalities that occurred during the F-84F Thunderbird era.

The Republic F-84F Thunderstreak would be in US service from 1950 through 1971. A total of 2,348 aircraft were built for the USAF and for foreign countries. *Republic Archives via LIRAHS*

F-84F Thunderbirds are in the diamond formation over the desert. There is a subtle difference between the lead aircraft and the other three, in that the lead has a red guide line on the leading edge of the wing. *USAF*

F-84F Thunderbirds are in the diamond formation and emitting smoke. The F-84F became the first model for the Thunderbirds that used smoke as part of the show. *USAF*

Maj. Jack Broughton, team leader, is preparing for a flight in 1955 in his F-84F Thunderstreak #1 aircraft, with his crew chief on the wing. Note that the flag display of different countries visited by the team is located directly on the side of the cockpit of the aircraft. In a three-year period, the Thunderbirds went to twelve different countries in Latin America. *Walt Jeffries via Mike Machat collection*

Rear three-quarter view of Maj. Broughton's F-84F with Jack standing at left. Broughton was the only Thunderbird Lead to have flown three different team aircraft: Republic's F-84G Thunderjet and F-84F Thunderstreak, then the supersonic North American F-100C Super Sabre. *Walt Jeffries via Mike Machaat Collection*

F-84F Thunderstreak, serial number 52-6583, in its original Thunderbirds paint scheme. It is on display at the PIMA Aerospace Museum in Tucson, Arizona. *Ken Neubeck*

F-84F Thunderbird aircraft ride down the runway after landing and with drag chutes deployed. *USAF*

THEN: In 1955, the F-84-flying Thunderbirds visited Republic for special performances before employees. Mundy Peale picked up team after each show and drove them past cheering crowds. From left, Ed Palmgren (soon to be Thunderbird Commander), Peale, Jack Broughton, Bill Creech and Billy Ellis.

Well-known aviation artist Mike Machat created this painting of four F-84F Thunderbirds aircraft participating in a special performance at the Republic Aviation plant in Farmingdale, New York, on Sunday, August 28, 1955, in front of thousands of Republic Aviation employees and their families. The event was to thank the employees of Republic for their contributions for the F-84F Thunderstreak production program. The painting was made in September 1997 to commemorate the fiftieth anniversary of the US Air Force. The original painting is in a private collection. *Mike Machat*

CHAPTER 3
North American
F-100C Super Sabre

F-100C SPECIFICATIONS

Wingspan	38 feet, 9.5 inches
Length	49 feet, 6 inches
Height	16 feet, 2.75 inches
Empty weight	20,450 pounds
Power plant	one Pratt & Whitney J57-7 turbojet engine
Maximum speed	864 mph
Service ceiling	45,000 feet
Range	1,500 miles

The team switched from the F-84F to the F-100C Super Sabre in June 1956, to be able to fly supersonic jet aircraft. Concurrently, the base for the team was moved from Luke AFB in Arizona to Nellis AFB near Las Vegas, Nevada, because of logistics concerns.

The team would begin to vary its maneuvers with the expanded performance capabilities of supersonic aircraft. The team would now have two additional aircraft added to do solo maneuvers. The six-aircraft delta formation became a regular maneuver to accompany the four-aircraft diamond formation for shows.

There would be five crashes during the team's use of the F-100C aircraft, with five fatalities occurring between 1957 and 1961. The team used the F-100C through the end of 1963.

A total of 476 F-100C Super Sabres were built for the US Air Force. The F-100C served with the Thunderbirds team from 1956 through 1963. *USAF*

Four F-100C Super Sabre Thunderbirds are flying in the diamond formation over Hoover Dam ca. 1955. *USAF*

F-100C Thunderbirds are flying in the six-aircraft delta formation, with the nose of another F-100 aircraft in the foreground. The introduction of this aircraft to the team formalized the addition of the two solo pilots into the air show routine and the use of the delta formation. *USAF*

F-100C Thunderbird, serial number 55-2723, is seen here on tarmac. *USAF via Michael Benolkin*

The first group of F-100C Super Sabre Thunderbirds have just been delivered to the team and are seen here taking off in the four-aircraft delta during practice in the Nevada desert, near Las Vegas. *USAF via Lou Drendel*

This team photo of the USAF Thunderbirds was taken in 1961 with Canada's Golden Hawks aerobatic team in front of the Thunderbirds F-100C aircraft. The Golden Hawks flew F-86 Sabres at that time, until 1965. *USAF via LIRAHS*

A group of F-100Cs are on the tarmac during an air show. At this time, the team did not use any type of aircraft numbering on the tail section, instead using the serial number. *USAF via Thunderbird Alumni Association*

A group of F-100Cs are on the tarmac during an air show. Of interest is the fact that there are eight F-100C aircraft that were used for this show, for which solo aircraft were added to the basic four-aircraft team. Also, it is noted that the #4 slot aircraft (located fourth from the left) features an uncleaned tail section from accumulated exhaust debris due to its position in the diamond. The aircraft would gain the nickname of Black Tail. *USAF via Thunderbird Alumni Association*

F-100C Thunderbirds are flying in inverted position during practice over the Nevada desert near Las Vegas, where the team relocated from Luke AFB in 1956. *USAF*

CHAPTER 4
Republic F-105B Thunderchief

F-105B SPECIFICATIONS	
Wingspan	34 feet, 11 inches
Length	64 feet, 3 inches
Height	19 feet, 8 inches
Empty weight	27,500 pounds
Power plant	one Pratt & Whitney J75-P-19 turbojet engine
Maximum speed	1,254 mph
Service ceiling	52,000 feet
Range	2,390 miles

For the 1964 air show season, the Thunderbirds made a decision to switch from the F-100C Super Sabre aircraft to a modified F-105B Thunderchief aircraft. The F-105 Thunderchief, at 50,000 pounds of maximum weight at takeoff, was the largest, heaviest and most powerful single-engine, single-seat fighter ever built. Using the F-105 aircraft would allow the Thunderbird team to be the first aerial team to reach Mach 2 speed. A total of nine modified F-105B aircraft were ordered, and they were pulled from active service. The aircraft were sent to Republic to be modified, and all were delivered to the team by the spring of 1964. Modifications included the removal of the APN-105 Doppler navigation system, the M-61 cannon, and associated fire control equipment, along with the APS-54 radar-warning system—items not needed for an air show aircraft.

The F-105B Thunderbirds began flying in April but unfortunately, on May 8, 1964, during the seventh show of the season at Hamilton AFB in California, one of the modified F-105B (serial number 57-5801) broke apart during a climb during the show, killing the pilot, Capt. Gene Devlin.

The F-105B Thunderbirds were then grounded while an investigation was conducted, which subsequently revealed that a structure splice plate in the top part of the fuselage failed due to fatigue. This would result in Project Backbone, during which a redesigned splice plate was retrofitted into all B models.

Some consideration was given to the use of modified F-105B aircraft for the Thunderbird team for the next year, but at that point the Air Force elected to use the F-100D Super Sabres to finish the season, as well as to be used for the next. The remaining eight Thunderbird F-105B aircraft were modified back to original

status and transferred to the 141st Tactical Fighter Squadron (TFS) of the New Jersey Air National Guard (ANG). Thus ended the shortest stint of any aircraft type that was used by the Thunderbirds. It would be the last Republic-built aircraft used by the team.

A total of seventy-one pruduction F-105B model Thunderchiefs were built. The B model was assigned to the Air National Guard units, while the F-105D model would be used heavily by USAF in bombing missions over North Vietnam during the Vietnam War, where significant amounts of aircraft were lost, leading to loss of operational capability. *USAF via LIRAHS*

The F-105B center console used for the Thunderbirds had modifications to make it compatible with the Thunderbirds' aircraft format. Controls for the gun and the APS-54 radar-warning system were either removed or deactivated. There were new controls that were added for the smoke release system that is used during the show. Also, located in the far left of the console is a control panel that is in the shape of a thunderbird logo, which controls special flight control functions for the F-105B Thunderbird aircraft. *Republic Archives via Cradle of Aviation*

This close-up view of the left side of the console shows the special thunderbird-shape panel that controls the 4-degree flap and knife-edge controls required for certain maneuvers that are specific for the Thunderbirds' air show. *Republic Archives via Cradle of Aviation*

A F-105B Thunderbird, serial number 57-5782, is being rolled out at Republic Airport in Farmingdale, New York, after the Republic Aviation Corporation modified the aircraft into the Thunderbirds configuration, during early 1964. Modifications included the removal of the gun and armament controls. *Republic Archives via Cradle of Aviation*

A F-105B Thunderbird, serial number 57-5782, is in flight over the Atlantic Ocean near Long Island after it was modified by the Republic Aviation Corporation in 1964. The smoke modification included a set of pipes located at the lower part of the exhaust section, used for putting the special mix of paraffin-based oil into the *exhaust* nozzle that creates smoke during an air show performance. *Republic Archives via Cradle of Aviation*

The underside of an F-105B Thunderbird, serial number 57-5782, in flight shows the distinct blue Thunderbirds paint scheme on the lower fuselage and the wings. The F-105B was the first Thunderbird model in which this particular paint scheme was introduced. Two external fuel tanks are loaded on the outboard wing stations for long-distance trips. *Republic Archives via Cradle of Aviation*

This is the Thunderbird team in the beginning of 1964 at their training location in Nevada, posing in front of the F-105B with team leader Maj. Paul Kauttu, shown on the far left. The accident that claimed the life of Capt. Gene Devlin (*second from the right*) resulted in the team ceasing to use the F-105B aircraft. *Thunderbird Museum via John Gourley*

This is the Thunderbird team as led by Maj. Paul Kauttu (*left*) in the latter part of 1964, posing in front of the F-100D Super Sabre aircraft that was brought in for them to fly as a replacement for the F-105B Thunderchief aircraft. *Thunderbird Museum via John Gourley*

CHAPTER 5

North American F-100D Super Sabre

F-100D SPECIFICATIONS

Wingspan	38 feet, 9.5 inches
Length	49 feet, 6 inches
Height	16 feet, 3 inches
Empty weight	21,000 pounds
Power plant	one Pratt & Whitney J57-20 turbojet engine
Maximum speed	864 mph
Service ceiling	45,000 feet
Range	1,500 miles

With the structural-failure issues adversely affecting the F-105B Thunderbirds aircraft, the Air Force decided not to use the F-105B anymore, and it had to find a suitable replacement quickly for the team. The decision was to go back to using the F-100 Super Sabre for the balance of the 1964 season, although it would be the upgraded D model, which was an improvement over the F-100C with regard to strengthening the structure and other reliability improvements.

The F-100D was an improvement over the F-100C with regard to improvements made to structure as well as with the reliability.

The F-100D Thunderbirds performed the most shows in the history of the team, with 121 shows in twenty-three countries in Europe, Latin America, and the Caribbean, as well as the United States. One highlight in 1965 was a nonstop team flight from Paris to Colorado that included seven in-flight refuelings.

In 1967, the team performed the thousandth show in its history, and in 1968 it received official status as an aerobatic USAF squadron.

There were two accidents involving the F-100D model during Thunderbird service, with the loss of two F-105D and one F-105F aircraft. The aircraft was used by the team until 1968.

F-100D Super Sabre aircraft is in action in Vietnam in 1967, firing rockets. A total of 1,274 F-100D aircraft were built by North American and served with the US Air Force from 1956 through 1979, when it was retired from ANG service. *USAF*

The F-100D Thunderbirds are in the four-aircraft delta formation during takeoff. The distinctive refueling probe that extends from the right wing can be seen on the F-100D aircraft. *USAF via Lou Drendel*

Thunderbirds F-100Ds in the classic six-aircraft delta formation in flight. The distinctive Thunderbird paint scheme on the lower fuselage and wings of the F-100D continues the tradition started from the previous F-105B Thunderbirds. *USAF via Lou Drendel*

This is the ill-fated two-seat F-100F Super Sabre Thunderbird, serial number 56-3924, that was flown by Maj. Frank Liethen and Capt. Robert Morgan. On October 12, 1966, this aircraft collided with another F-105D aircraft during a training flight near Indian Springs, Nevada, while performing the Cuban Eight maneuver. Both crew members of the F-100F aircraft died during the subsequent crash, while the other aircraft involved, the F-100D, was able to land successfully. *USAF via Lou Drendel*

This F-100D Super Sabre Thunderbird #6, serial number 55-3520, seen here in early 1967, was flown by Merrill "Tony" McPeak and would be lost in an accident. During an air show at Laughlin AFB, Texas, on October 20, 1967, McPeak was conducting a vertical rolling climb during the bomb burst maneuver when both wings sheared off from the fuselage. McPeak was able to eject successfully, and he would have a successful career in the US Air Force, eventually reaching the rank of four-star general in 1988. Failure analysis revealed that there was a weakness in the wing center box of the F-100D, where overstress during flight would cause the wing to shear off. Subsequently, this would lead to a 4 g limit being placed on all Thunderbird maneuvers during air shows. *USAF via Lou Drendel*

F-100D Thunderbird #5 is coming in for a landing after performing at an air show in Oxnard, California, in 1968. In contrast to when the F-100C flew with the team, the individual team member aircraft would have the aircraft team number marked on the tail section of the aircraft in lieu of the serial number. *USAF via Thunderbird association*

Thunderbird F-100D #6 is on display at the US Air Force National Museum in Dayton, Ohio. This aircraft served with the team from 1964 through 1968 and then returned to active USAF duty until it was retired from the South Dakota ANG, in 1977. *John Gourley*

With the advent of the F-100D in Thunderbirds service, the F-100D Thunderbird #4 slot aircraft featured a black tail from being in the rear position, which gets the full impact of engine exhaust products. The maintenance crew has left the tail uncleaned to make this the black-tail Thunderbird. *USAF via Thunderbird Alumni Association*

F-100D Thunderbird #5 is on a tarmac in an unidentified airfield and is the second of two solo aircraft that was added to the team. Note that the #4 slot aircraft is to the left and has a dirty or "black tail" from engine exhaust soot, which covers the original red-and-white paint scheme on the tail. *USAF via Thunderbird Alumni Association*

USAF officers and dignitaries are viewing the solo aircraft of the F-100D team perform maneuvers at Randolph AFB, Texas, in April 1968. *USAF via Lou Drendel*

F-100D Thunderbirds are performing a flyover for the benefit of employees of the North American Aviation plant in California, where the aircraft were produced. *USAF*

CHAPTER 6

McDonnell F-4E Phantom II

F-4E SPECIFICATIONS

Wingspan	38 feet, 5 inches
Length	63 feet
Height	16 feet, 3 inches
Empty weight	29,000 pounds
Power plant	two GE J79-15 turbojet engines
Maximum speed	1,473 mph
Service ceiling	60,000 feet
Range	1,245 miles

The Thunderbirds switched to the F-4E Phantom in 1969, which, at Mach 2, was significantly faster than the F-100D aircraft.

The year 1969 began with the team still flying the F-100D aircraft, and an accident occurred during training with one of the solo aircraft. It was at this point that the team decided to go with just one solo pilot when they converted to the F-4E Phantom, which would be continued for the next few years due to the ongoing fuel crisis in the US. Thus, a number of maneuvers such as the six-aircraft delta would now be a five-aircraft formation.

The F-4 was the most powerful aircraft flown by the Thunderbirds at this point; it was fast, big, and heavy. The F-4 Phantom would be the only aircraft model that would be used both by the Thunderbirds and the Blue Angels (who flew the F-4J from 1969 through 1973).

The F-4 Phantom was almost double the speed of its predecessor, F-100D, and this allowed for more capabilities during air show performances. The F-4E could quickly go into maneuvers, with increased speed and capabilities compared to the F-100D.

During the F-4E years there were two accidents, with the loss of two aircraft and three lives. By 1973, both the Thunderbirds and the Blue Angels were affected by public opinion regarding the military and the ongoing fuel crisis in the US. The Thunderbirds were particularly affected by the crisis during 1973; the team performed only six shows in the US and was grounded for the rest of the time.

The McDonnell F-4 aircraft first flew in 1959 and served for both the US Navy and USAF. It saw significant action in the Vietnam War. Over 5,195 F-4 aircraft were built, with some of them still in foreign service. *USAF photo by MSgt. Jim Wines*

With the deletion of one solo pilot from the Thunderbird team, the F-4E Phantom Thunderbirds would fly in a five-aircraft delta formation in lieu of the six-aircraft delta formation flown by previous teams, as seen here during the team performing maneuvers in Puerto Rico. Note that in this photo and others of the F-4 Thunderbird team, the #4 aircraft has a black tail. *USAF via Lou Drendel*

The F-4 Phantom Thunderbirds are seen here doing a landing-gear-down maneuver, which slows down four of the five aircraft in the five-aircraft delta formation during flight while the #5 aircraft passes underneath, during an air show in Duluth, Minnesota, in August 1970. *USAF via Lou Drendel*

This Thunderbirds F-4E aircraft is completing a turn maneuver during an air show in June of 1971. This would be the first model that had twin engines that the team would fly. *USAF via Lou Drendel*

Five-aircraft Thunderbirds F-4E formation in flight. The paint scheme was modified during transition to F-4 from F-100 model, including painting of a large number of the aircraft number on the vertical tail section. *USAF via Lou Drendel*

This schematic diagram comes from the USAF flight-operating difference/supplemental-data manual for the F-4E Thunderbirds, and it shows how the smoke-generating system is controlled by the pilot for the F-4E Thunderbird aircraft. In contrast to smoke systems used on previous Thunderbird aircraft models, the smoke oil was stored in the dummy missiles that were mounted at the aft and forward stations. In addition, it can be seen from the diagram that the smoke system was pressurized to ensure reliability of the system during negative and inverted maneuvers.

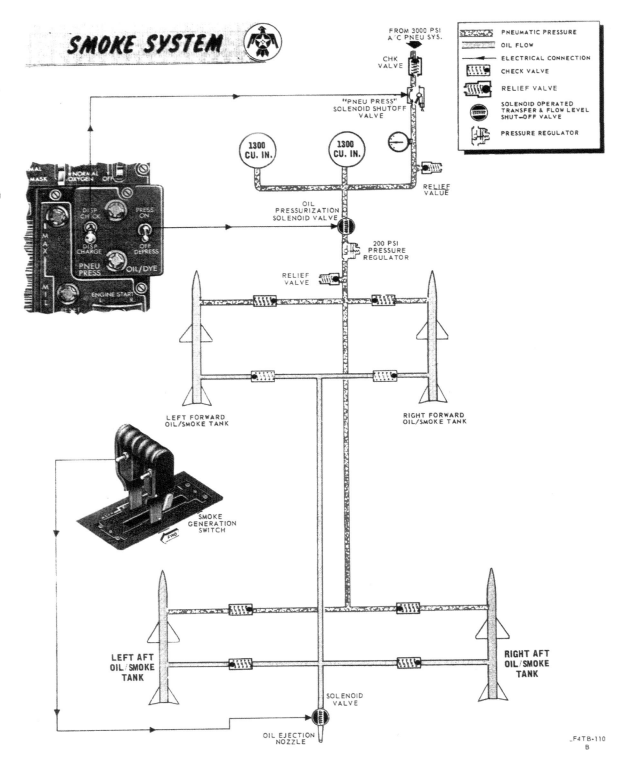

SMOKE SYSTEM

	PNEUMATIC PRESSURE
	OIL FLOW
	ELECTRICAL CONNECTION
	CHECK VALVE
	RELIEF VALVE
	SOLENOID OPERATED TRANSFER & FLOW LEVEL SHUT-OFF VALVE
	PRESSURE REGULATOR

FROM 3000 PSI A/C PNEU SYS.

CHK VALVE

"PNEU PRESS" SOLENOID SHUTOFF VALVE

1300 CU. IN.

1300 CU. IN.

RELIEF VALVE

OIL PRESSURIZATION SOLENOID VALVE

200 PSI PRESSURE REGULATOR

RELIEF VALVE

DISP CHECK

PRESS ON

DISP CHARGE

OFF DEPRESS

PNEU PRESS

OIL/DYE

ENGINE START

NORMAL / OXYGEN / OFF

LEFT FORWARD OIL/SMOKE TANK

RIGHT FORWARD OIL/SMOKE TANK

SMOKE GENERATION SWITCH

FWD

LEFT AFT OIL/SMOKE TANK

RIGHT AFT OIL/SMOKE TANK

SOLENOID VALVE

OIL EJECTION NOZZLE

_F4TB-110 B

The F-4E Thunderbird team waits on a wet tarmac during an air show appearance. *USAF via Lou Drendel*

When the F-4 Phantom Thunderbirds perform the five-aircraft delta formation, the #4 slot aircraft (with the black tail) is riding on the right side, as seen during this pass over a coastal city. *USAF via Lou Drendel*

This is the five-man Thunderbird team in 1970 at their training location in Nevada, posing in front of the F-4E, with team leader Lt. Col. Tom Swalm shown in the center. *Thunderbird Museum via John Gourley*

This is a view of the back-seat rider while two F-4E jets pass from behind. Note that the crew member is wearing a regular uniform and not a G suit. *USAF via Lou Drendel*

The team is performing a vertical-dive maneuver during an air show, demonstrating the raw power of the F-4E Phantom aircraft. *USAF photo by R. J. Archer via Lou Drendel*

This is a front view of the F-4E Thunderbird #1 aircraft, and it shows the specific paint scheme used by the team on this aircraft. In addition to the large number of the aircraft painted on the vertical tail section, the aircraft number is painted on the lower forward fuselage, in front of the nose landing-gear door. This aircraft has two external fuel tanks located under the wings to aid in traveling extended distances to the next air show location. *USAF via Lou Drendel*

The addition of another aircraft to the team during the F-4E Phantom era has allowed for more formations to be developed by the team. Two different formations are demonstrated in these photos of the team practicing at Nellis AFB. The photo on the left shows the team in a seven-aircraft modified delta formation in the form of a arrowhead formation, with the seventh aircraft riding in the rear, whereas the photo on the right is the standard diamond formation. *USAF via Lou Drendel*

This caricature of the F-4E Thunderbirds flight-operating supplemental manual, showing the Birds in delta formation, pays tribute to the Native American tradition embraced by the team.

CHAPTER 7

Northrop T-38A Talon

T-38 SPECIFICATIONS

Wingspan	25 feet, 3 inches
Length	45 feet, 4.5 inches
Height	12 feet, 10.5 inches
Empty weight	7,200 pounds
Power plant	two GE J85-5A turbojet engines
Maximum speed	858 mph
Service ceiling	50,000 feet
Range	1,140 miles

In 1974, the team would switch from the F-4E to the T-38 as a result of the fuel crisis in the US. The T-38A was not a frontline jet fighter like the previous models used by the team, but it had sufficient capabilities to be used by the team. Most important of all, it was more economical with regard to fuel: five T-38 Talon aircraft used the same amount of fuel needed for one F-4 Phantom aircraft.

The switch to the T-38 also saw an alteration of the flight routine to exhibit the aircraft's maneuverability in tight turns. It also ended the era of using the black tail on the #4 slot aircraft, which would have the tail cleaned and be just like the other aircraft in the team.

During the T-38 era, there were four crashes that involved the loss of seven aircraft, but the worst crash in the team's history occurred on January 18, 1982, during training at Indian Springs, Nevada. While practicing the four-plane diamond loop, the four-aircraft formation impacted the ground at high speed, instantly killing all four pilots: Maj. Norman L. Lowry, Capt. Willie Mays, Capt. Joseph N. Peterson, and Capt. Mark Melancon.

The cause of the crash was determined by the USAF to be the result of a mechanical problem with the #1 aircraft's control stick actuator in the trail flap, which caused insufficient back pressure on the T-38 control stick during the loop. Since the rest of the team would visually cue off the lead aircraft during formation maneuvering, the team disregarded their positions relative to the ground and followed the lead aircraft into the ground.

The January 1982 crash ended the use of the T-38 Talon by the Thunderbirds, as flying was suspended for the rest of the year pending investigation of the crash. It was announced later in 1982 that the Thunderbirds would continue to fly as a team, but with the newer F-16 fighter jets beginning in 1983.

The T-38 aircraft first flew in 1959 with the USAF. Over 1,146 T-38s were built, the F-5 being a variant of this aircraft. *USAF*

View of the Thunderbirds T-38 #1 aircraft shows that the number is in a circle in the middle of the tail, similar to the previous paint scheme used on the F-4 Phantom II Thunderbird aircraft. Since the T-38 is a trainer aircraft, all aircraft are two seats in tandem. *USAF via Lou Drendel*

The #4 aircraft is undergoing ground maintenance during an air show appearance. The T-38 was a lighter and smaller aircraft than the F-4 Phantom II, by a difference of over 20,000 pounds in empty weight and 500 mph less in maximum speed. The Talon used much less fuel than the team's previous aircraft, the F-4E Phantom. *USAF via Lou Drendel*

Thunderbird #1 through #7 are lined up for a publicity photo on a tarmac prior to an air show. The aircraft have the canopy open for this shot. *USAF via Lou Drendel*

The paint scheme for the T-38 Thunderbirds used a scallop-design scheme on the nose, and wide lines on the wings and tail section on the lower fuselage and wings, in lieu of the thunderbird shape used on the F-4E Thunderbirds. *USAF via Lou Drendel*

T-38 Talon aircraft are in the trail formation over an airfield, with the first three aircraft being viewed from the cockpit of the #4 aircraft view. *USAF via Lou Drendel*

This view from the back seat of a T-38 Talon cockpit shows it following another aircraft while banking, in a trail maneuver where all of the aircraft are in serial formation while in flying in different orientations. *USAF via Lou Drendel*

This view inside of a T-38 Thunderbird Talon cockpit is looking to the rear of an observer in the rear seat while the aircraft is sandwiched between the #2 and #3 aircraft during takeoff. *Dan Wilbur*

The same camera setup from above shows the aircraft and the #2 and #3 aircraft during a vertical climb. *Dan Wilbur*

T-38 Thunderbird #5 is taking off with both pilot and back-seat passenger during an air show in June 1980. *USAF photo by SSgt. Charles Diggs*

T-38 Thunderbirds are performing an arrowhead loop during an air show in June 1980. *USAF photo by SSgt. Charles Diggs*

This view from a T-38 Talon cockpit shows it banking with the other aircraft in the team, including the #2 aircraft on the right, while over an airfield. The T-38 Thunderbird did not have in-flight refueling capabilities, so the team was restricted from air shows on the other side of the ocean. *USAF via Lou Drendel*

The T-38 Thunderbird team is flying in a six-aircraft delta formation as it passes over the New Jersey shore, ca. 1980. With the T-38, the GE turbojet engines ran clean so that the #4 slot aircraft's tail would remain clean, thus bringing an end to the "Black Tail" aircraft. Note that there is a back-seat crew member in all aircraft during this flight. *Dan Wilbur*

The #5 aircraft is performing a climb during an air show. The pilot is located in the front seat, and the rear seat may be occupied by a maintenance crew member or may be left empty during performances. *USAF via Lou Drendel*

Bottom view of the four-aircraft T-38 diamond formation. The color scheme is a pattern using stripes. *USAF via Lou Drendel*

Similar to what was done for the F-4E, a schematic diagram was presented in the USAF flight-operating difference/supplemental-data manual for the T-38 Thunderbirds, and it shows how the smoke-generating system is controlled by the pilot for T-38 Thunderbird aircraft, in which the smoke oil is released on the left engine tailpipe.

SMOKE SYSTEM

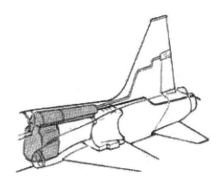

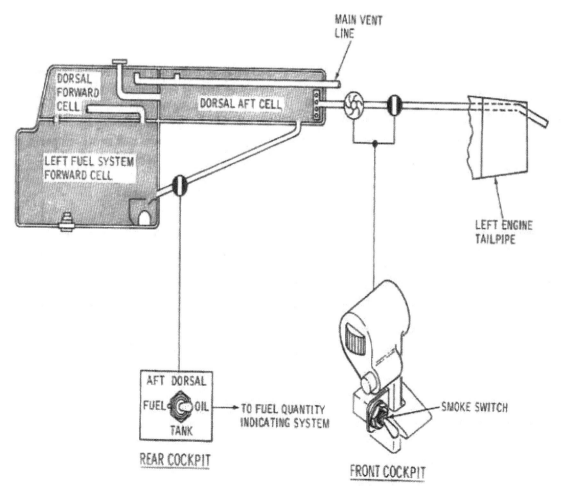

MAIN VENT LINE

DORSAL FORWARD CELL

DORSAL AFT CELL

LEFT FUEL SYSTEM FORWARD CELL

LEFT ENGINE TAILPIPE

AFT DORSAL

FUEL — OIL

TANK

→ TO FUEL QUANTITY INDICATING SYSTEM

REAR COCKPIT

SMOKE SWITCH

FRONT COCKPIT

As is the custom of other aerobatic teams, publicity photos are taken with the USAF T-38 Thunderbirds with US national monuments and landmarks in the background. Here is a photo of the five-aircraft T-38 Thunderbirds team next to the arches of St. Louis, Missouri. *USAF via Lou Drendel*

Here is a photo of the five-aircraft T-38 Thunderbirds team flying past Mount Rushmore, South Dakota. *USAF via Lou Drendel*

Here is a photo of the five-aircraft T-38 Thunderbirds team flying pas the Statue of Liberty in New York City. *USAF via Lou Drendel*

Here is another photo opportunity for the five-aircraft T-38 Thunderbirds team as they fly past the Golden Gate Bridge in San Francisco, California. *USAF via Lou Drendel*

A four-aircraft T-38 diamond formation has just come out from a loop at an air show in Wilmington, North Carolina, in 1980. *USAF photo by Bill Stephenson*

There is a static display of the T-38 Thunderbird #1 aircraft at the US Air Force Academy at Colorado Springs, as seen here during the Waldo Canyon wildfire in June 2015. *USAF photo by Mike Kaplan*

CHAPTER 8
Lockheed F-16 Fighting Falcon

F-16C SPECIFICATIONS

Wingspan	32 feet, 8 inches
Length	49 feet, 5 inches
Height	16 feet
Empty weight	18,900 pounds
Power plant	one Pratt & Whitney F100-PW-200 turbojet, or one GE F110-GE-129 turbojet
Maximum speed	1,500 mph
Service ceiling	50,000 feet
Range (ferry)	2,002 miles

In 1983, the Thunderbirds resumed flying with the F-16 Fighting Falcon, after the tragic crash of four T-38 aircraft in 1982. With the transition to the F-16A Fighting Falcon aircraft, the team returned to using a frontline fighter.

Beginning in June 1982, the transition to the F-16 aircraft for the Thunderbirds team was led by Maj. Jim Latham. The first public air show was held on April 2, 1983, at Nellis AFB, Nevada.

In 1992, the team would transition from the F-16A and the two-seat B model to the F-16C and two-seat D model. For the C and D models, there were upgrades to the avionics and radar that improved the performance of the aircraft over the previous model.

The Thunderbirds use the F-16C and D models of aircraft up to the current day.

The F-16 Fighting Falcon has been the staple of the USAF fighter force. Over 4,600 aircraft have been produced for the US and for many other countries around the world. *USAF photo by MSgt. Don Taggart*

After the T-38 Diamond Crash tragedy in 1982, the Thunderbirds prepared to resume flying as a team with a new aircraft, the F-16A Fighting Falcon, in April 1983. Here are the team's maintainers working on the F-16A for the first time in preparation for the team's first air show in Nevada. *USAF photo by SSgt. Bob Simons / National Archives*

F-16A Thunderbird #5 in flight during the Nellis AFB air show. Of interest in this photo is that the #5 that is painted on the lower fuselage is in the upright orientation, compared to the inverted-5 configuration that will be used in a few years for this aircraft. *USAF photo by SSgt. Bob Simons / National Archives*

A major event for the team was achieved when they resumed flying after the T-38 Diamond Crash, in their first air show in front of the public with F-16A aircraft at Nellis AFB, in April 1983. Here is the team as led by Maj. Jim Lathan in the #1 aircraft in the diamond formation. *USAF photo by SSgt. Bob Simons / National Archives*

The paint scheme for the Thunderbird F-16A Fighting Falcon follows what was used for the F-4E Thunderbird aircraft used previously by the team, with the distinctive bird shape on the lower fuselage and wings. The team also continued to use smoke as part of the show, as seen here. *USAF photo by SSgt. Bob Simons / National Archives*

This two-photo sequence shows the team performing the arrowhead roll as one of the maneuvers conducted during the first public show at Nellis AFB, in April 1983. *USAF photo by SSgt. Bob Simons / National Archives*

F-16A Thunderbirds are in six-aircraft delta formation over Lake Mead in Nevada, ca. 1983.

F-16A Thunderbirds #5 and #7 are performing maneuvers during the open-house air show at George AFB, California, in 1989. This photo is interesting in that the #7 aircraft is performing a solo maneuver for this show, and the #5 aircraft now has the inverted #5 located on its fuselage. *USAF photo by MSgt. Patrick Nugent*

The F-16A Thunderbirds are flying in the six-aircraft delta formation at the mountains near Nellis AFB in November 1983, concluding a successful year for the Thunderbirds' return to active performances in the US. It is interesting to note that even with the faster F-16 aircraft model compared to the previous T-38 aircraft model (at almost double the speed), the #4 slot aircraft did not experience the effects of the "black tail" experienced on the F-100 and F-4, since the engine exhaust ran cleaner. *USAF photo by SSgt. Bill Thompson / National Archives*

One of the original F-16A Fighting Falcon aircraft, serial number 81-0663, is on display at the Museum of Aviation at Warner Robbins Air Force, Georgia. *John Gourley*

The F-16A model was used by the Thunderbirds from 1983 through 1992, until the team transitioned to the F-16C model, which had upgraded avionics in the cockpit as compared to the older model. *John Gourley*

The F-16C Fighting Falcon aircraft that were introduced to the team in 1992 (from the F-16A version) have been modified in several areas from the Air Force version of the F-16C. One is that the M61 Vulcan machine gun and ammo drum in the nose cone have been replaced with a special tank that holds smoke oil. The paraffin-based smoke oil in the tank is injected into the plane's exhaust nozzles, instantly vaporizing it, creating the smoke trails used in the show, as seen in the maneuver by the team that is shown here prior to the landing sequence. *Ken Neubeck*

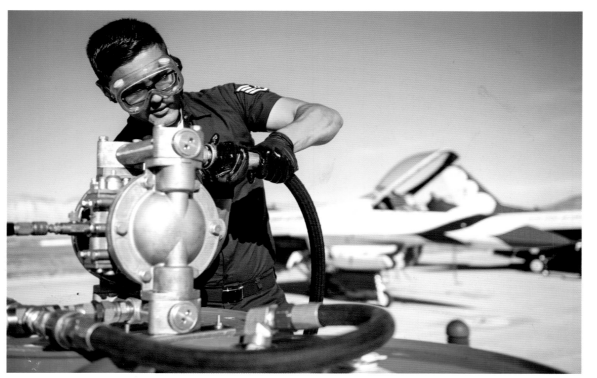

Maintenance personnel filling the special tank in the aircraft with the smoke oil from a drum that will be used to create smoke for an upcoming Thunderbird performance in 2015. *USAF photo by Senior Airman Rachel Maxwell*

This F-16C cockpit is open, allowing a view of the console section in the front and the ACES II ejection seat located in the rear. *Ken Neubeck*

The F-16C cockpit contains a number of displays. The glass screen for projections of the heads-up display (HUD) is on the top of the console. The control panel below the HUD glass is the integrated control panel (ICP). Located to the left of the ICP is a clock with a timer. *Ken Neubeck*

The F-16 aircraft is the most powerful aircraft ever flown by the Thunderbirds. The aircraft is powered by a single Pratt & Whitney F100 turbofan engine. *Ken Neubeck*

The nose landing gear for the F-16 is in extended position. *Ken Neubeck*

This is the P&W F100 turbofan engine being tested. The F100 engine is used on both the F-15 and F-16 aircraft. *USAF photo by SSgt. Robert Zoellner*

These are the main landing gear of the F-16 aircraft, each gear having two links and an actuator. *Ken Neubeck*

Another special feature of the Thunderbirds is that all aircraft on the team have a dedicated square on the fuselage that shows the flags of the different countries that the Thunderbirds have visited in Europe and Latin America. This feature goes back over sixty years, to when the team used the F-84F Thunderstreak aircraft.
Ken Neubeck

Another view of the flag emblem display location, in relation to the canopy area of the F-16C aircraft.
Ken Neubeck

For some trips, the team may elect to carry an external fuel tank underneath the fuselage to minimize the use of in-flight refueling. Here is the team's F-16D aircraft equipped with an external 360-gallon tank mounted on the centerline of the lower fuselage.
Ken Neubeck

This is a closeup view of the external fuel tank that is mounted on the centerline of the lower fuselage of the F16D aircraft.
Ken Neubeck

For some trips, particularly cross-country trips or overseas trips to a foreign country, aerial refueling will be required. Here is the Thunderbird team being refueled by a KC-135 tanker. *USAF photo by TSgt. Christopher Boitz*

A Thunderbirds F-16C jet pulls away after refueling from a KC-135 tanker from the 319th Air Refueling Wing above Grand Forks AFB, North Dakota, in October 2007. *USAF photo by Airman 1st Class Chad M. Kellum*

Pilot Alex Turner in the Number 5 aircraft gives thumbs-up indication during taxiing at the TICO air show held in Titusville, Florida, in March 2017. *John Gourley*

Three aircraft perform the start of the breakaway maneuver during the TICO air show in Titusville, Florida, in March 2017. *John Gourley*

F-16 Thunderbird #4 flown by Maj. Whit Collins lands at Long Island MacArthur Airport in the May 2019, for the Jones Beach air show, after traveling from the last air show deployment at Kirkland AFB in New Mexico. Note that the aircraft lands on the rear wheels first during the landing sequence. *Ken Neubeck*

An F-16 Thunderbird aircraft takes off to participate with the team during the Jones Beach Air Show in May 2019. *Ken Neubeck*

F-16 Thunderbird #1, during landing sequence at MacArthur Airport on Long Island in May 2019, is flown by the team leader, Lt. Col. John Caldwell. The #1 aircraft is often referred to as the "Boss." *Ken Neubeck*

F-16 Thunderbird #2, flown by Maj. Will Graeff, follows a few seconds behind the #1 during a serial landing sequence on Long Island during May 2019. *Ken Neubeck*

This F-16 Thunderbird #3 aircraft is flown by Capt. Michael Brewer and is coming in for a landing after an air show performance on Long Island in May 2019. *Ken Neubeck*

F-16 Thunderbird #4, flown by Maj. Whit Collins, is coming in for a landing near runway lights at MacArthur Airport after a practice performance on Long Island, May 2019. The speed brakes, located on both sides of the engine exhaust, are fully deployed during this landing. *Ken Neubeck*

Thunderbird #5 lead solo aircraft is flown by Maj. Alex Turner during an air show performance for the TICO air show in Titusville, Florida, in March 2017. The #5 aircraft maintains the tradition of the inverted "5" painted on the fuselage to signify that this aircraft does many of the inverted maneuvers during an air show. *John Gourley*

Thunderbird #6 opposing solo aircraft is flown by Capt. Michelle Curran and is coming in for a landing on Long Island in May 2018 after a practice run for the Jones Beach Air Show. The outline of the thunderbird wing design can be seen on the lower wing surface. *Ken Neubeck*

The F-16 Thunderbird tail section contains movable surfaces on the vertical tail section and on the inside of the horizontal section that act like speed brakes.
John Gourley

F-16 Thunderbird aircraft prepare for a practice run at Republic Airport on Long Island in May 2009. The vertical tail section is being moved on the aircraft while the pilot is steering the aircraft.
Ken Neubeck

This is the F-16 aircraft used by the Thunderbirds team during a trip to Long Island at Republic Airport in 2009 for the Jones Beach Air Show. This would be the last year that the team would use Republic Airport as its base for the air show, since the runway length is not quite long enough for the safety protocol required by the F-16 Thunderbirds aircraft (a waiver was issued by the USAF for the Thunderbirds to use this airfield for 2009). *Ken Neubeck*

This is the F-16 Thunderbird team during a trip to Long Island at MacArthur Airport ten years later, in May 2019, for the Jones Beach Air Show. The team has been using this airport for the Jones Beach Air Show in recent years because of the longer runway available, to meet the team's safety protocol. *Ken Neubeck*

The F-16 Thunderbirds are doing a delta formation pass during a photo opportunity by the Golden Gate Bridge after performing at an air show in California on their way back to home base in Nevada in late September 2018. *USAF photo by SSgt. Ashley Corkins*

The F-16 Thunderbirds are doing a pass-in-review formation near the Freedom Tower in lower Manhattan during a photo opportunity over New York City in early September 2018. The different cities that are visited by the Thunderbirds are among those visited by other US aero teams such as the Blue Angels. *USAF photo by SSgt. Ned T. Johnston*

The four-aircraft F-16 Thunderbirds team has just created a heart-shaped symbol in the skies over Cannon AFB, New Mexico, in May 2018. *USAF photo by Senior Airman Luke Kitterman*

The F-16 Thunderbirds are joined in formation by the Royal Canadian Snowbirds CT-114 trainer aircraft during an air show performance at the Air National Guard Base at Boise, Idaho, in October 2017. The Thunderbirds join with other teams on a yearly basis for photo opportunities. *ANG photo by MSgt. Joshua Allmaras*

This is another perspective of the four-aircraft diamond formation as performed by the team during an air show performance at Scott AFB, Illinois, in August 2006. *USAF photo by MSgt Jack Braden*

CHAPTER 9

Thunderbirds Support Aircraft and Crew

When the Thunderbirds were flying the F-84F Thunderchief model in 1955, they received their first transport support aircraft, which was a C-119 Flying Boxcar painted in team colors. This transport aircraft would eventually be complimented by the C-123 Provider aircraft to support the team. Both aircraft types would be painted in Thunderbird motif.

The support aircraft was not actually part of the air show, as is the case with the "Fat Albert" C-130 for the Blue Angels. If the base is near enough, the transport aircraft may fly back to its base, leaving the maintenance crew to work with the team during the air show.

The biggest tragedy suffered by the Thunderbirds team occurred on October 10, 1958, when the C-123D aircraft, serial number 55-4521, left Utah to head to Washington; en route, the aircraft suffered what appeared to be a bird strike over Payette, Idaho. All nineteen crew members were killed. A monument with the names of those who perished was built by the Payette Kiwanis and the high school's Key Club.

For the last three decades, both C-130 and C-17 transport aircraft have been assigned to the Thunderbirds from different USAF bases, depending on where the team flies from. The aircraft will bring typically sixty to seventy maintenance personnel to support the team for each air show event.

The C-119 Flying Boxcar would be one of the first dedicated transport aircraft for the USAF Thunderbirds that was used to carry spare parts and maintenance crew. Pictured here is C-119, serial number 51-8166, one of a few C-119s that would serve with the team from 1954 through 1959. This particular C-119 would be returned to USAF service and eventually be sold to Taiwan. *Bill Larkins*

The front section of C-119, serial number 51-6146, displays the Thunderbirds emblem and the Luke AFB designation on the front fuselage. This transport aircraft supported the Thunderbirds when the team was flying F-84 aircraft. *USAF*

The Thunderbirds would also use the C-123 Provider transport aircraft to support the team during the 1950s and 1960s. Seen here is C-123, tail number 54-0671. The tail section of this aircraft has been painted with the scallop design and stars, reflecting the paint scheme of the Thunderbird team. *Bill Larkins*

The Thunderbirds have used a variety of USAF transport aircraft during their history, including the time with the F-16 aircraft. Here is the F-16A Thunderbirds team in delta formation following their assigned transport, a C-141B Starlifter, in early 1983.
National Archives

Maintenance crew stand on top of the C-130 support aircraft, watching the Thunderbirds perform during an event in Guatemala City in October 2005.
USAF photo by SSgt. Matthew Lohr

Maintenance crew members for the Thunderbirds team are unloading equipment off an Air Force C-17 aircraft shortly after it had landed at Long Island's MacArthur Airport. This particular C-17 was deployed from Dover AFB, Delaware, and was assigned to the Thunderbirds for the Long Island air show in May 2019. The C-17 has a large rear cargo door that is ideal for equipment to be unloaded by tractors and forklift vehicles. *Ken Neubeck*

A C-17 aircraft has just landed at Long Island MacArthur Airport from Dover AFB, Delaware, on the Monday after an air show weekend to pick up crew and equipment. As the team does not have a dedicated transport aircraft like the Blue Angels team have with the Fat Albert C-130, arrangements have to be made for a USAF transport aircraft such as C-130 or C-17, from an active base for each air show event. *Ken Neubeck*

Maintenance crew chiefs for the T-38 Thunderbirds team are running toward their assigned aircraft during preflight activities at an air show in 1980. *USAF photo by TSgt. John L. Marine*

Maintenance crew chiefs for the Thunderbirds team are departing a C-17 aircraft shortly after arrival at McChord AFB, Washington, for the Skyfest air show in May 2014. The support aircraft arrives at the designated air show airport first, prior to the arrival of the Thunderbird jets. *USAF photo by TSgt. Sean Tobin*

Dozens of maintainers and other support personnel have left the C-17 transport at MacArthur Airport for support of the Thunderbirds team during their visit to Long Island in May 2019 for the Jones Beach Air Show. Crew uniforms are dark blue and have the Thunderbird patch. *Ken Neubeck*

A Thunderbirds maintainer is using a military forklift to carry maintainers' personnel luggage unloaded from the transport aircraft. *Ken Neubeck*

A maintainer is unloading a large tool and parts locker from the transport aircraft. The transport aircraft is completely unloaded for each air show event and then leaves. *Ken Neubeck*

A key piece of equipment that the Thunderbirds carry for each show is a spare Pratt & Whitney F100 engine on its own transport cart. The crew has the capability of swapping out an engine on the F-16 aircraft if necessary. *Ken Neubeck*

This two-seat F-16D aircraft was equipped with three external fuel tanks after making a long trip across the United States. *Ken Neubeck*

A maintainer is replenishing hydraulic fluids into the hydraulic system of the #5 aircraft. *Ken Neubeck*

Crew remove one of the 370-gallon wing external fuel tanks from F-16D aircraft to a storage area. *Ken Neubeck*

Maintenance personnel are doing some repairs in the cockpit area of the aircraft. *Ken Neubeck*

Maintenance support personnel is waiting for his assigned aircraft to arrive after conducting a practice run of the Thunderbird #1 aircraft at Republic Airport, Long Island, in May 2009. *Ken Neubeck*

The #2 aircraft has arrived and is turning into its designated slot on the tarmac while crew chiefs remain in position. *Ken Neubeck*

Thunderbird #3 is receiving instructions from the crew chief shortly after landing in May 2009.
Ken Neubeck

Thunderbird #2 aircraft is going through preflight check, with the maintenance personnel observing from the rear of the aircraft.
Ken Neubeck

A maintainer is working in the cockpit area during preparation of the two-seat F-16D aircraft for upcoming media flights during a visit to Long Island in 2019. Note the portable ladder assembly, used to access the cockpit. *Ken Neubeck*

Another view of the maintainer working on the two-seat F-16D aircraft, using the portable ladder assembly. *Ken Neubeck*

CHAPTER 10
Thunderbirds Media Aircraft

Media flights became an important part of a Thunderbirds visit during an air show. The first media aircraft that was used by the team was the T-33 Shooting Star. As the Thunderbirds progressed during the 1950s, it was found that a two-seat aircraft would be beneficial to perform several secondary tasks for the team, such as carrying the narrator for the team as well as flying members of the media during an air show event. Because the Thunderjet was a single-seat fighter, a two-seat T-33 Shooting Star served as the narrator's aircraft and was used as the VIP/press ride aircraft. The T-33 served with the Thunderbirds in this capacity in the 1950s and 1960s.

When the team flew the two-seat T-38 Talon beginning in 1974, media flights were conducted in the #8 aircraft in the team.

The current process for the F-16 Thunderbirds team is to conduct a maximum of two media flights during a visit, with the flights being performed on the afternoon or morning after the team's arrival. One of the flights may be assigned for a hometown hero to ride in the back seat. The two-seat F-16D aircraft is used for these flights, and it may either be the #7 aircraft (if it is a two-seat F-16D aircraft) or the #8, depending on aircraft assignment. In some cases, a decal with the back-seat media person's name will be placed on the side of the cockpit.

The role of media aircraft became an integral part of the team with the introduction of T-33 Shooting Star trainer aircraft, serial number 53-5547, to be used by the team's narrator. Here is the aircraft at Hamilton Field in March 1959. *Bill Larkins*

Thunderbird T-38 #8 is at Republic Airport on Long Island supporting an air show in the 1980s. This aircraft was used for the media flights that were conducted during this air show. *Republic Archive via LIRAHS*

This F-16D two-seat aircraft was flown by the advance pilot / narrator of the team and arrived the day before the main team arrived. As can be seen, there is no number yet inserted in the circle that is located on the lower fuselage in preparation for this aircraft being used for media flights. In this case during a visit to Long Island, New York, in May 2017, the #8 will be used, since the other numbers have been assigned to other seven aircraft of the team. *Ken Neubeck*

In this photo a two-seat F-16D aircraft that was originally used as the #8 advance aircraft is being prepared by the maintenance crew for transforming it into the media aircraft during an air show on Long Island in May 2019. The aircraft will be used to fly media members in the back seat during an air show event. Note that this aircraft has had the #8 decal removed from inside of the circle below the cockpit, and the pilots' names on the side of the cockpit have also been removed.
Ken Neubeck

In the photo on the left, the two-seat F-16 aircraft is being shown by the ground crew to members of the media during an arrival event of the team, while a maintainer is applying a number. The new decal used is #7, as seen in the photo on the right, and this aircraft will be used for the media flights for this air show.
Ken Neubeck

The rear station of the F-16D two-seat aircraft is a full crew station that is equipped with ACES II ejection seat. It has its own separate boarding ladder. *Ken Neubeck*

The rear station of the F-16D has two multifunction displays, similar to what is in the front station of the aircraft. *Ken Neubeck*

Media flight at Shaw AFB, South Carolina, is conducted with a local newscaster sitting in the back seat of F-16D #7, in May 2012. *USAF photo by Airman 1st Class Hunter Brady*

An NHL hockey player from the St. Louis Blues is in the back seat for the media flight held at Scott AFB, Illinois, in September 2012. *USAF photo by SSgt. Brian J. Valencia*

The airport where the Thunderbirds team is based for an air show is a major local media attraction, not only for the show performances themselves but also during the practice days. Here are spectators and photographers looking upward at the aircraft smoke trail on both sides of the airport fence at Republic Airport, Long Island, in May 2009.
Ken Neubeck

The base from where the team works during an air show presents a number of photo opportunities. The two-seat F-16D media aircraft (no number assigned yet) is stationary on the ground, while F-16C Thunderbird #1 flies over the Republic Airport on Long Island in preparation for landing.
Ken Neubeck

Nighttime photo of the F-16 Thunderbirds team at the TICO air show in Florida, 2017, shows the media aircraft (#8) on the outside of the eight-aircraft team for this visit. *John Gourley*

CHAPTER 11
Thunderbirds Pilot Selection and Training

The 1967 Thunderbird F-100D team, led by their team leader, Maj. Neil Eddins. It is noted on this photo that the team number of the aircraft is seen here on the front of the nose landing-gear door, as was done until the team used the F-4E Phantom. *Thunderbird Museum via John Gourley*

Thunderbird pilots are very experienced fliers, with many of the pilots having operational experience with the F-16 as well as trainer instructor experience. Up to the present time, there is a rigorous selection process for the individual pilots on the Thunderbirds team.

Each year, three demonstration pilots are changed in the team. All candidates must have at least 1,000 flying hours accrued on a jet fighter, as well as having been in the Air Force for at least three years. Candidates must be current with the F-16 aircraft.

The process involves candidates submitting an application to the team, on the basis of which semifinalists are selected. After selection, four to eight prospective team members visit the Thunderbird's hangar at Nellis AFB, where each candidate participates in an evaluation flight in the back seat of the F-16D two-seat aircraft. During the flight, the front-seat pilot will perform formation flying and other fighter aircraft maneuvers. The leader evaluates the finalists and sends recommendations through the chain up to Air Combat Command, prior to final selections.

As part of the training, there is a three-week training course that allows the new members to integrate better into the team. The team has used Las Vegas, Nevada, as their winter training base for over fifty years, since the weather there is ideal for winter training. Training goes from November through March, and at the end the team is ready for the new air show season.

Service with the Thunderbirds is typically for two years; sometimes for three, if the pilot serves with the media aircraft initially. After service, the pilot will return to active duty in the US Air Force.

Maj. Sean Gustafson, the pilot for Thunderbird #4 (slot), is leaving his aircraft just after it has arrived at Republic Airport on Long Island in preparation for the Jones Beach Air Show in May 2009. Maj. Gustafson would be the first Air Force reserve pilot to join the Thunderbirds. *Ken Neubeck*

Pilots from Thunderbirds #1 through #6 are conferring with other team members shortly after arrival at Republic Airport on Long Island in May 2009. They will soon be conducting media interviews next to their aircraft. *Ken Neubeck*

Thunderbird #7 F-16C aircraft, piloted by Lt. Col. Kevin Walsh, arrives on the runway at MacArthur Airport under poor weather conditions on Long Island in May 2017. Capt. Walsh was a native Long Islander who returned to his hometown during this visit for the annual air show. The #7 aircraft is the operational officer, responsible for administration functions for the team. *Ken Neubeck*

Capt. Walsh conducts media interviews in front of his #7 aircraft in May 2017. In 2018, he would be promoted to the team leader, flying the #1 aircraft. *Ken Neubeck*

Maj. Whit Collins is seen here shortly after arrival being interviewed by members of the local media next to his Number 4 F-16C aircraft at Long Island MacArthur Airport in May 2019. At the time, he was entering his third year of service on the team and had previously served as the Number 6 solo pilot. Each team member is assigned a number of local media people during the pilot interview session. *Ken Neubeck*

Maj. Whit Collins is flying the #4 aircraft during a landing during a return to MacArthur Airport after performing at the Jones Beach Air Show, Long Island, in May 2019. *Ken Neubeck*

Capt. Michelle Curran is landing the #6 opposing solo aircraft at MacArthur Airport after an air show performance on Long Island in 2019. This was her first season with the Thunderbirds team. *Ken Neubeck*

Capt. Michelle Curran, #6 pilot, is being interviewed by local media during the Thunderbirds arrival event in May 2019 on Long Island. Her #6 aircraft has been towed so that it's in the background for the interview. *Ken Neubeck*

Maj. Alex Turner, #5 opposing solo pilot, is landing at Long Island's MacArthur Airport during misty rain and fog conditions prior to an air show event to be conducted at Jones Beach the following days in May 2017. The team is ready for any kind of adverse weather conditions at any time and at any location that they may encounter. *Ken Neubeck*

Maj. Alex Turner, #5 aircraft pilot, is photographed in front of his aircraft during the media event during the team's arrival in May 2017 on Long Island. In June 2016, Maj. Turner, in Thunderbird #6, crashed in a field near Colorado Springs after performing a flyover at the US Air Force Academy graduation ceremony. He safely ejected and was unhurt. Investigation revealed that the aircraft's engine was inadvertently shut down at the start of landing procedures due to a faulty throttle trigger rotating to an engine cutoff position. *Ken Neubeck*

In 2005, Maj. Nicole Malachowski would become the first woman pilot for the Thunderbirds team, when she flew the #3 (right wing) aircraft. She is seen here during a practice run at . She would serve with the team until the summer of 2007. *USAF photo by SSgt. Kristi Machado*

In 2006, Maj. Samantha Weeks would become the second woman pilot for the Thunderbirds team, when she flew in the #6 (opposing solo) aircraft. She is seen here meeting with the public at the air show at Offut AFB, Nebraska, in August 2008. She would serve with the team until the summer of 2008. *USAF photo by Lance Cheung*

This is a painting that portrays the four Thunderbird T-38 pilots who perished during in the "Diamond Crash" on January 18, 1982, during training in Las Vegas, Nevada. The officers were the commander, Maj. Norman L. Lowry; Capt. Willie Mays (left wing); Capt. Joseph ("Pete") N. Peterson (right wing); and Capt. Mark Melancon (slot pilot). *Painting by Jeanette Pajares, USAF Art Collection / National Archives*

Thunderbirds team leader Lt. Col. Kevin Walsh calls roll call prior to the playing of "Taps" at the funeral ceremony held for Maj. Stephen Del Bagno at his high school, Saugus High School in Santa Clarita, California, on April 15, 2018. Maj. Del Bagno lost his life in an accident during training at Nellis in April 2018. *USAF photo by SSgt. Stephanie Englar*

Two F-16s (including one Thunderbird) and two F-35s fly over during the funeral service for Maj. Stephen Del Bagno, held in Santa Clarita, California. These were two of the thirty different aircraft models that were flown by Bagno during his USAF career. *USAF photo by SSgt. Stephanie Englar*

On occasion at some locations, the commander of the Thunderbirds will be asked to perform swearing-in ceremonies for new Air Force recruits. Here, Lt. Col. John R. Venable prepares to swear in new recruits for the Nebraska ANG during an air show appearance by the team in April 2001 at Lincoln Municipal Airport, Nebraska. *USAF photo by SSgt. Matthew A. Rolan*

Lt. Col. Richard G. McSpadden signs autographs for the crowd at the open house of Eglin AFB, Florida, in April 2002. The team does find opportunities to meet with the public in between flying. *USAF photo by Joe Piccorossi (civilian)*

Capt. Christopher Stricklin ejects from his F-16 during an air show at Mountain Home AFB, Idaho, in September 2003. The aircraft was in the process of performing a split-S maneuver when a miscalculation in the altitude occurred, due to the difference in the height between Mountain Home's location (3,000 ft.) and the team's home base in Nellis AFB, Nevada (2,000 ft.). Procedures were changed where the mean sea level (MSL) altitude readings would be used for the proper altitude for the procedure in lieu of the above ground level (AGL) altitude. *USAF photo by SSgt. Bennie J. Davis III*

CHAPTER 12

Thunderbirds Air Show Maneuvers

The Thunderbirds in the basic six-aircraft delta maneuver over the Hudson River in New York City in August 2019. *Ken Neubeck*

The Thunderbirds average about sixty shows a year at thirty different locations. Thus there is a need to have the same well-defined set of routines and maneuvers in order to maintain consistency and, most of all, safety. The Thunderbirds practice these routines during the off season at Las Vegas, Nevada, where the team will practice six days a week and accumulate a total of 120 training missions.

There are three different sets of routines—the High show, the Low show, and the Flat show. The selection of which show will be performed on the day of the air show depends on different factors, such as the weather, cloud ceiling, and prevailing winds. The following table shows the setup for these shows:

NAME	CEILING	UNIQUE MANEUVERS
High show	8,000 feet	40
Low show	3,500 feet	10
Flat show	2,000 feet	6

As has been established over the years with previous Thunderbirds teams, there are specific job titles for each of the different members/aircraft of the Thunderbirds team:

#1	Team Leader (Boss)
#2	Right Wing
#3	Left Wing
#4	Slot
#5	Lead Solo
#6	Opposing Solo
#7	Operations Officer
#8	Advance Pilot and Narrator

Thunderbirds #1 through #4 perform the delta formation, with the two solo aircraft, #5 and #6, performing around them for certain maneuvers as well as individual maneuvers away from the team. All six aircraft do perform together in the delta-style maneuvers.

The following pages show some of the maneuvers in pictorial form, along with photos illustrating how the maneuvers are performed during the show. Most team maneuvers come out initially from either the diamond or the delta formation. The maneuver descriptions come from the *Thunderbirds Maneuvers Manual*.

Getting a basic understanding of and being able to identify these maneuvers add to the enjoyment of watching the show, as does taking photos.

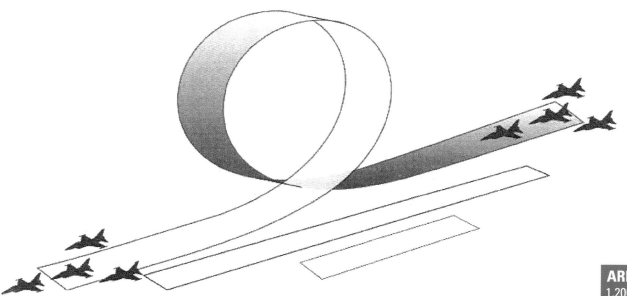

Arrowhead Loop

Maneuver: The four-aircraft arrowhead formation will come in at crowd right at 100 ft. AGL, at 1,200–1,500 ft. from the crowd, and perform a loop at CP (crowd center) and proceed toward crowd left. **Used in the High show.**

The four-aircraft arrowhead formation performs the arrowhead loop during a practice performance at Joint Base Andrews, Maryland, in September 2015. *USAF photo by Senior Airman Jason Coulilard*

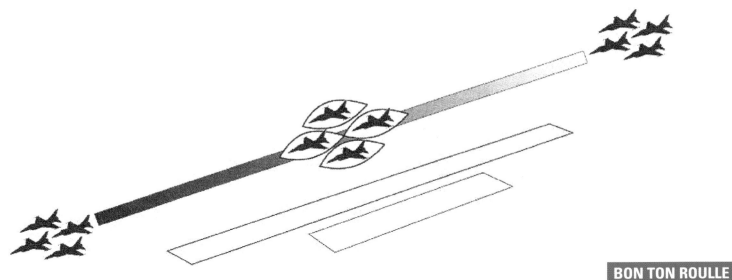

Bon Ton Roulle

Maneuver: The four-aircraft diamond formation will come in at crowd right at 100 ft. AGL, at 1,200–1,500 ft. from the crowd, and all four aircraft will simultaneously invert at CP (crowd center) and proceed toward crowd left. **Used in the High show.**

Thunderbirds #1 through #4 perform the Bon Ton Roulle maneuver, remaining in the four-aircraft diamond formation while all four aircraft simultaneously invert at the center point above the crowd, during an air show at Keesler AFB, Mississippi, in March 2015. *USAF photo by TSgt. Manuel Martinez*

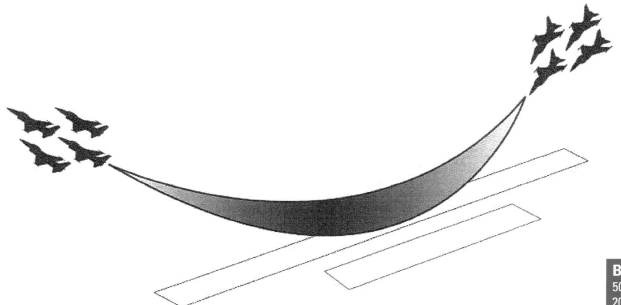

Bottom-Up Pass
Maneuver: The four-aircraft diamond formation will initially perform a descending pass at 200 ft. AGL on the 500 ft. show line from crowd right and ascending to crowd left. **Used in the Low show.**

The four-aircraft diamond formation performs an ascending maneuver during the bottom-up pass at a practice performance for an air show held at Lakeland, Florida, in April 2015. *USAF photo by TSgt. Manuel Martinez*

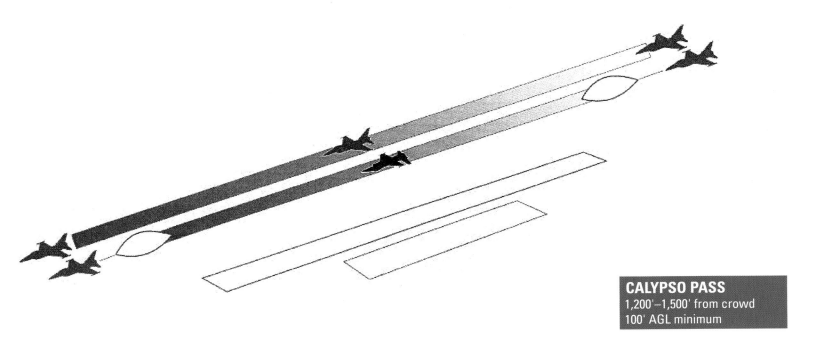

Calypso Pass

Maneuver: The two solo aircraft will ingress from the right of the crowd, with #5 rolling into inverted position above the #6 aircraft, at 100 ft. AGL on the 1,200–1,500 ft. show line, and exit to crowd right. **Used in the High show.**

Thunderbird aircraft #5 and #6 are performing the Calypso Pass with #5 inverted while performing over the crowds at an air show held at Dyess AFB, Texas, in April 2010. It is noted that the #5 on the aircraft is normally inverted and now appears normal. *USAF photo by Senior Airman Stephen Reyes*

CROSSOVER BREAK
1,200'–1,500' from crowd
500' AGL minimum

Crossover Break

Maneuver: The two solo aircraft enter from behind the crowd at 500 ft. AGL and then cross over above the crowd, with one aircraft going left and the other going right. **Used in the High show.**

Thunderbird solo aircraft #5 (lead solo flown by Maj. Rick Goodman) and #6 (opposing solo flown by Capt. Aaron Jelinek) perform the crossover break maneuver over the crowds at an air show held at Dyess AFB, Texas, in April 2010. *USAF photo by Senior Airman Stephen Reyes*

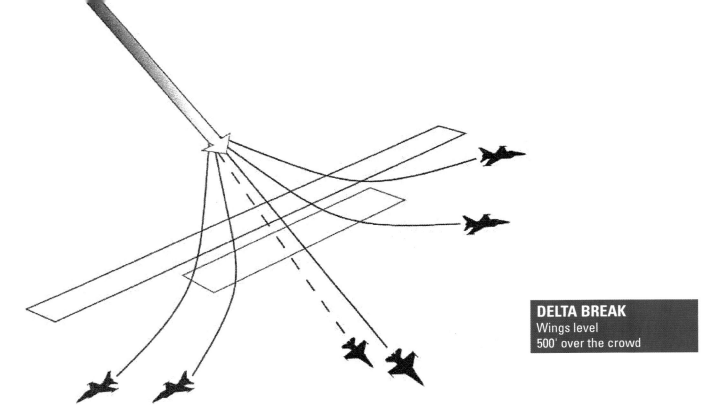

Delta Break

Maneuver: The delta will roll out in a head-on setup and commence a climb toward 500 ft. over the crowd. Aircraft #1 will pull up into a steady climb straight ahead. Aircraft #5 and #6 will pull outboard in a steady, 60-degree angle of bank, offset by 90 degrees. Then, #2 and #3 will pull outboard in a steady, 45-degree angle of bank, offset by 45 degrees. Finally, #1 and #4 continue straight ahead and exit behind the crowd. **Used in the High show.**

The Thunderbirds perform a delta burst maneuver in November 2018 in Homestead, Florida. The #1 aircraft is at center at the top, pulling up, and the #4 aircraft is at center at the bottom, going straight ahead. *USAF photo by MSgt. Mark Olsen*

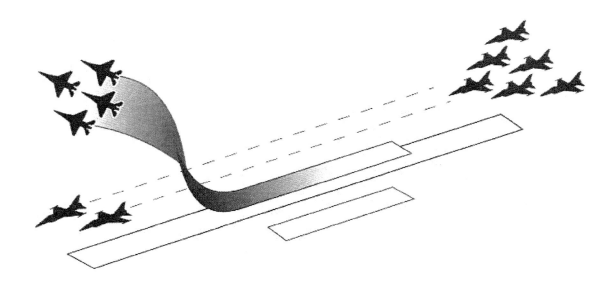

Delta Flat Pass

Maneuver: Ingressing straight and level at 200 ft. AGL from crowd right, all six aircraft will perform a flat pass on the 500 ft. show line and exit in front of the crowd to set up for the blue-out or pitch-up. **Used in the High, Low, and Flat shows.**

Here is an outside view of the Thunderbird team in classic delta flat pass formation, complete with smoke over Long Island's MacArthur Airport on their return from their air show performance at Jones Beach in May 2019. From this formation, the team will perform a blue-out or pitch-up, which results in each aircraft landing individually in serial fashion. *Ken Neubeck*

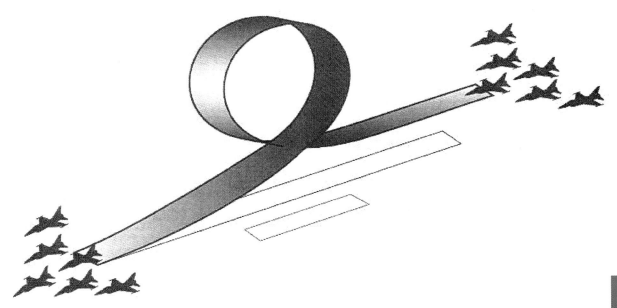

Delta Loop

Maneuver: The delta will ingress from the right of the crowd at 200 ft. AGL and will conduct a 360-degree roll at CP and continue forward. The #1 and #4 aircraft continue straight ahead and exit behind the crowd. **Used in the High show.**

View of the Thunderbirds from underneath in the delta formation as they perform the Delta Loop at the Point Magu air show in California in March 2007. *USAF photo by TSgt. Justin D. Pyle*

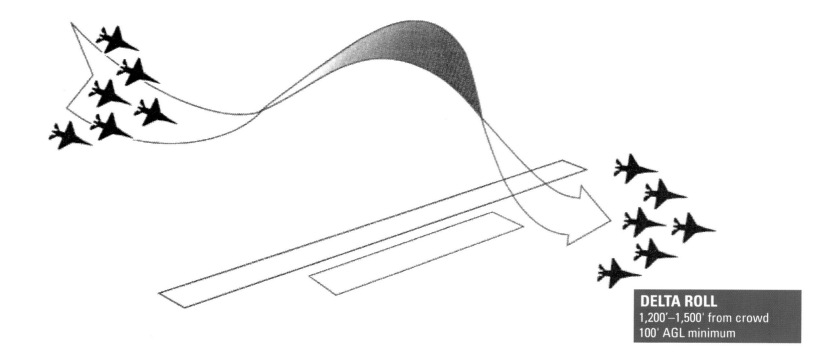

Delta Roll

Maneuver: Ingressing from altitude and reaching 100 ft. AGL from crowd left, six aircraft in delta formation will perform a roll at 1,200–1,500 ft. from the show line and exit to the right of the crowd. **Used in the High show.**

Thunderbird aircraft #1 through #6 are in delta formation while performing a roll over the crowds at an air show held at Dyess AFB, Texas, in April 2010. *USAF photo by Senior Airman Stephen Reyes*

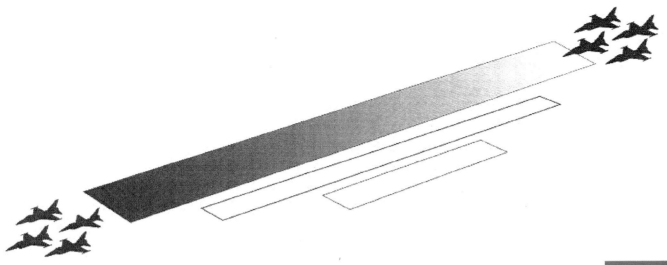

DIAMOND FLAT PASS
500' from crowd
200' AGL minimum

Diamond Flat Pass

Maneuver: Ingressing straight and level at 200 ft. AGL from crowd right, four aircraft in diamond formation will perform a flat pass on the 500 ft. show line and exit in front of the crowd. **Used in the High, Low, and Flat shows.**

Here is a view of the Thunderbird team in classic diamond flat pass formation, consisting of aircraft #1 through #4 at the air show held at TICO in Florida in 2017.
John Gourley

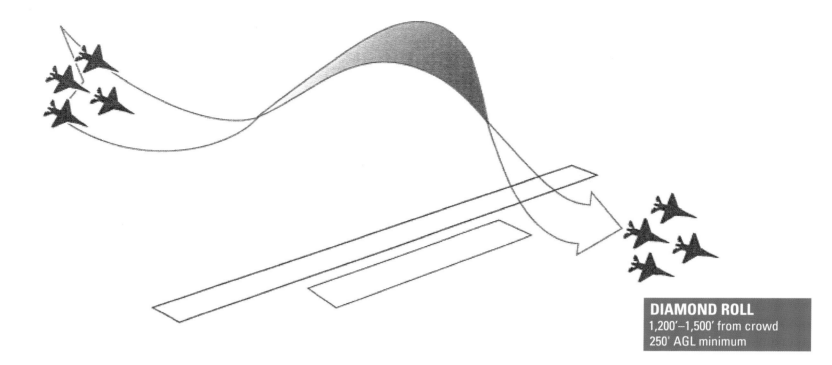

DIAMOND ROLL
1,200'–1,500' from crowd
250' AGL minimum

Diamond Roll

Maneuver: Ingressing from altitude at 250' AGL from crowd left, four aircraft in diamond formation will perform a roll at 1,200–1,500 ft. from the show line and exit to the right of the crowd. **Used in the High show.**

Thunderbird aircraft #1 through #4 are in the classic diamond formation and performing a roll during training over the Nevada test range in February 2018. *USAF photo by SSgt. Ned T. Johnston*

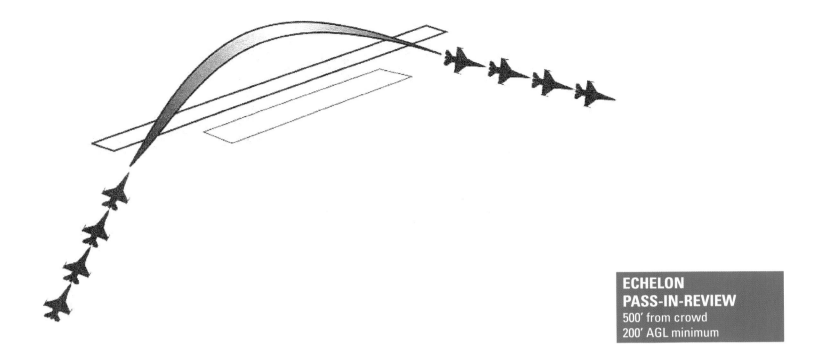

Echelon Pass-in-Review

Maneuver: Ingressing straight and level at 200 ft. AGL from crowd left, four aircraft in echelon formation will perform a flat pass on the 500 ft. show line and exit in front of the crowd. **Used in the Low show.**

Thunderbirds #1 through #4 fly in echelon formation as they pass over the crowd during an air show held at MacDill AFB, Florida, in April 2005. *USAF photo by TSgt. Sean Matteo White*

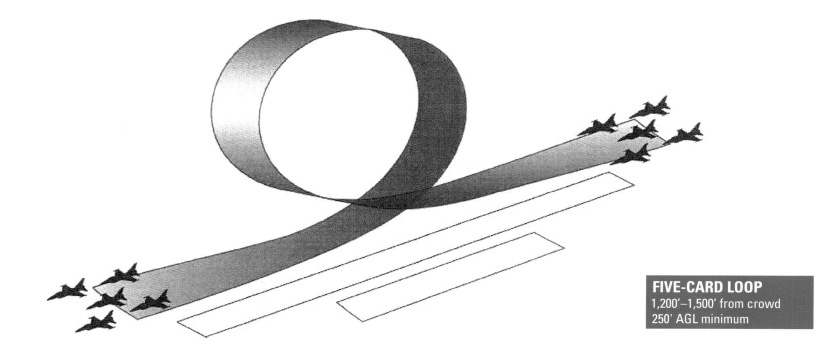

Five-Card Loop

Maneuver: Ingressing straight and level at 200 ft. AGL from crowd right, five aircraft will perform a coordinated loop at CP at the 1,200–1,500 ft. show line and exit to the left of the crowd. **Used in the High show.**

Thunderbirds #1 through #5 fly in five-card formation as they perform a loop over the crowd during an air show held at Travis AFB, California, in May 2014. The shadow of the moon can be seen in the background. *USAF photo by SSgt. Larry E. Reid Jr.*

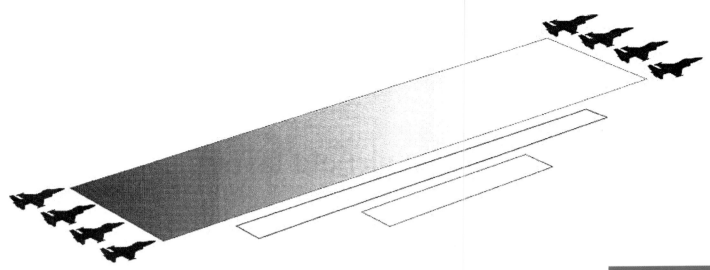

Line-Abreast
Flat-Pass

Maneuver: Ingressing straight and level at 200 ft. AGL from crowd left, four aircraft in line-abreast formation will perform a flat pass on the 500 ft. show line and exit in front of the crowd. **Used in the High, Low, and Flat shows.**

Thunderbirds #1 through #4 fly in line-abreast formation during an air show in Turku, Finland, in June 2017. *USAF photo by SSgt. Larry E. Reid Jr.*

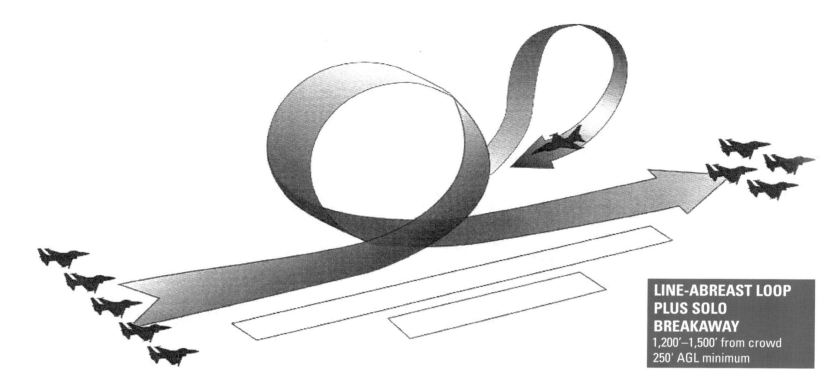

Line-Abreast Loop

Maneuver: Ingressing straight and level at 200 ft. AGL from crowd left, five aircraft in line-abreast formation will perform a loop at CP at 250 ft. AGL. The #5 aircraft will break away and head left while the four remaining aircraft go into diamond formation and exit crowd right. **Used in the High show.**

Thunderbirds in the five-aircraft line-abreast formation are performing a loop at the Joint Base McGuire-Dix-Lakeland (New Jersey) Air Show in May 2014. *USAF photo by SSgt. Larry E. Reid Jr.*

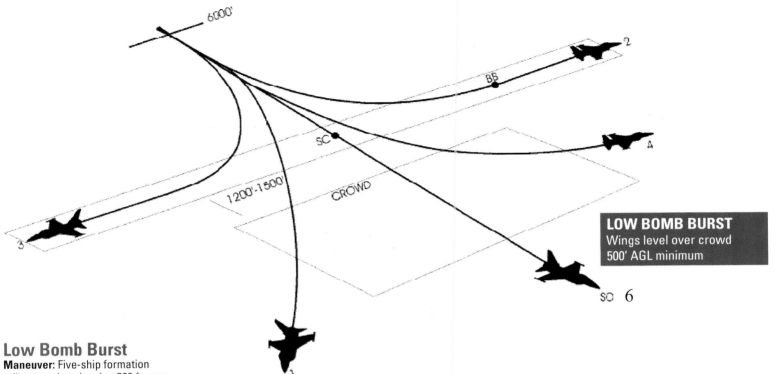

6000'

2

BB

4

SC

1200'-1500'

CROWD

3

LOW BOMB BURST
Wings level over crowd
500' AGL minimum

SC 6

1

Low Bomb Burst

Maneuver: Five-ship formation rollout at wings level at 500 ft. over the crowd. Aircraft #6 will pull straight ahead. The #2 and #3 aircraft will pull outboard in a steady, 2 g, 60-degree angle of bank, offset by 90 degrees. Aircraft #1 and #4 will pull outboard in a steady 2 g, 45-degree angle of bank, offset by 45 degrees. **Used in the Low show.**

Thunderbirds five-ship formation performs a low bomb burst maneuver in August 2006, during a practice performance at Scott AFB in Illinois. *USAF photo by MSgt. Jack Braden*

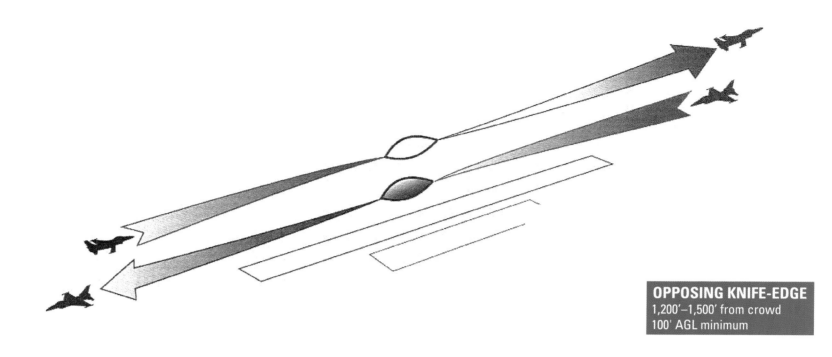

Opposing Knife-Edge

Maneuver: Solo aircraft #5 and #6 fly from opposite ends of the field at 100 ft. AGL at 1,200–1,500 ft. from the crowd. At CP each pilot will roll his aircraft into a 90-degree angle of bank prior to the cross. After the cross, both aircraft will roll upright and clear in front of the crowd. **Used in the High show.**

The two opposing solos, Thunderbirds #5 and #6, flown by Capt. Nicholas Eberling and Maj. Alex Turner, pass each other in the opposing knife-edge maneuver during the Luke AFB air show in April 2016, with F-35A static display aircraft in the foreground. *USAF photo by TSgt. Christopher Boitz*

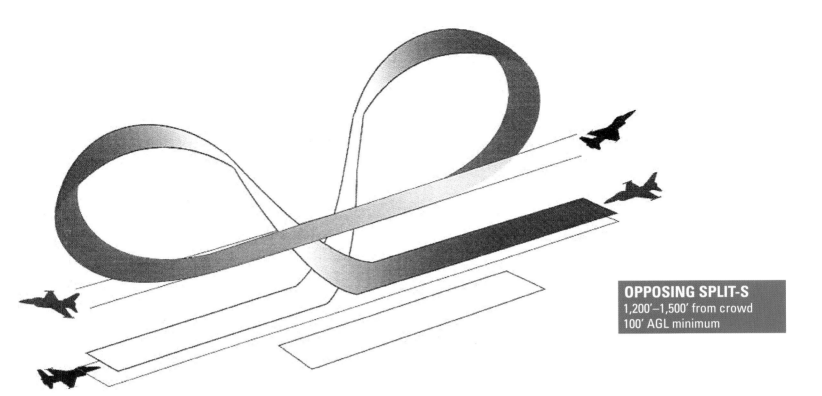

Opposing Split-S

Maneuver: Solo aircraft #5 and #6 fly from opposite ends of the field at 100 ft. AGL at 1,200–1,500 ft. from the crowd. At the point where the aircraft cross, each aircraft will pull up and perform a complete 360-degree loop and proceed straight. **Used in the High show.**

The two opposing solos, Thunderbirds #5 and #6, pass each other in the opposing split-S maneuver during the Charleston (South Carolina) AFB air show in April 2011, with F-35A static display aircraft in the foreground. *USAF photo by Airman 1st Class James Richardson*

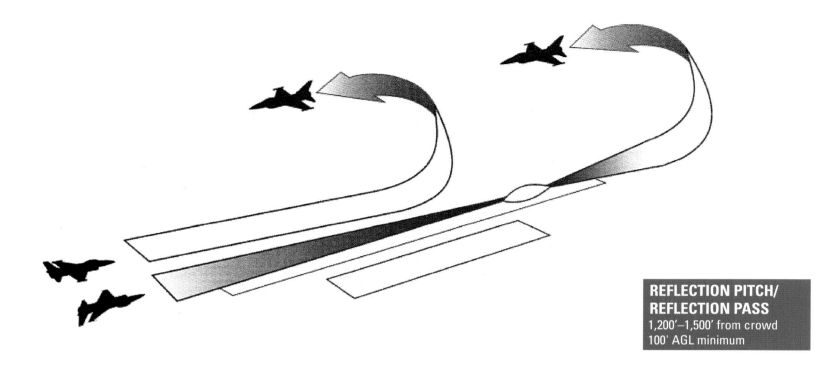

**REFLECTION PITCH/
REFLECTION PASS**
1,200'–1,500' from crowd
100' AGL minimum

Reflection Pass

Maneuver: Approaching CP from the left, the #5 and #6 aircraft fly at 100 ft. AGL, with #5 inverted and below the #6 aircraft. Each aircraft pulls 65–70 degrees nose up and then rolls 180 degrees. **Used in the High show.**

Thunderbirds solo aircraft #5 and #6 fly with #5 inverted below #6 during air show practice at Scott AFB, Illinois, in August 2006. The #5 on the aircraft can be seen in upright position. *USAF photo by MSgt. Jack Braden*

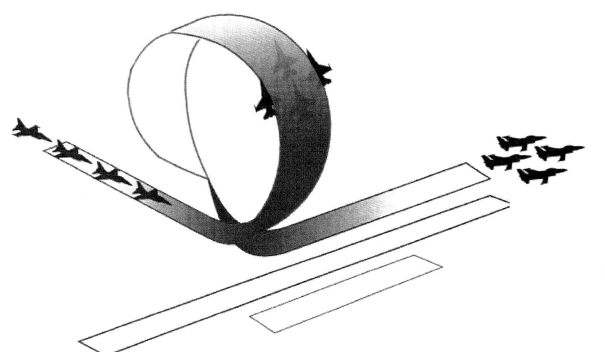

Trail-to-Diamond Clover Loop

Maneuver: Approaching from the front of the crowd, the four-aircraft trail formation reaches CP at 250 ft. AGL and performs a clover loop, then goes into a four-aircraft diamond formation and exits crowd right. **Used in the High show.**

The four-aircraft Thunderbird formation performs a trail-to-diamond clover loop maneuver during practice at Fort Worth, Texas, in October 2014. *USAF photo by TSgt. Manual J. Martinez*

Here are the #5 and #6 T-38 aircraft practicing the reflection pass maneuver in the Nevada mountains, with the #5 aircraft flying in the inverted position. *Dan Wilbur*

The T-38 six-aircraft delta formation during practice in the desert over Nevada, ca. 1980. *Dan Wilbur*

Thunderbirds #1 through #4 are in the process of performing an arrowhead loop maneuver during the final air show of the year at Nellis AFB, Nevada, in November 2009. *USAF photo by TSgt. Michael Holzworth*

Head-on view of the Thunderbirds team as they release smoke over Canon AFB, New Mexico, during a practice run in May 2018. *USAF photo by Senior Airman Luke Kitteman*

Head-on view of the Thunderbirds team during refueling from a KC-135 tanker during a trip from Nevada to Cleveland, Ohio, for an air show in September 2015. *USAF photo by Senior TSgt. Abigail Klein*

One of the more popular destinations for the Thunderbirds for several years has been the Atlantic City boardwalk in southern New Jersey. Here is the T-38 Thunderbirds team passing by the boardwalk area during the summer, ca. 1980. *Dan Wilbur*

In 2003, visits by the Thunderbirds to Atlantic City were formalized as an event as Thunder over the Boardwalk. Here is the F-16 Thunderbirds team flying low over the Atlantic City beach during the Thunder over the Boardwalk event in August 2011, to the thrill of spectators and photographers. The event occurs on a Wednesday, in contrast to the weekend days that are part of most air shows. *USAF photo by SSgt. Larry E. Reid Jr.*

CHAPTER 13

Thunderbirds Visits to Foreign Countries

The Thunderbirds have been visiting foreign countries since the team's inception in 1953. More than any US-based aerobatic team, the Thunderbirds live up to their reputation as America's Ambassadors in Blue by conducting regular visits to countries around the world.

This trend started with the F-84G Thunderbirds making trips to Brazil and other South American countries in 1954, and it continues up to the present day with the F-16 Thunderbirds. The visits are meant to foster good will, and, on occasion, the team works with other aerobatic teams such as the Red Arrows from the UK.

Members of the team such as the team leader will conduct interviews with the local media of each foreign country that is visited. Some historic visits by the team include a visit to mainland China in 1987 and to former Eastern Block nations in Europe during the 1990s and into the twenty-first century.

At the Thunderbirds Museum in Las Vegas, Nevada, there is a display of flags of the over sixty countries visited by the Thunderbirds team during their history. This display panel is similar to the display that is on the fuselage of the Thunderbirds aircraft. *John Gourley*

An early international trip for the F-16A Thunderbirds was participation at the Air Fete '84 show in June 1984 at RAF Mildenhall, England. The Thunderbirds are in the foreground, and the British Red Arrows team is located toward the top. Note that there is the basic six-aircraft team, along with the #7 and #8 F-16 Thunderbird aircraft. *USAF photo by TSgt. Jose Lopez Jr. / National Archives*

Members of the F-16A Thunderbird team receive preflight instructions from the team leader at RAF Upper Heyford in June 1984. *USAF photo by TSgt. J. Manson / National Archives*

In 1987, the Thunderbirds visited mainland China during their Pacific tour. Here is Capt. Bert Nelson, the narrator for the team, reviewing his script for the show with a Chinese translator at the Beijing Air Base. *USAF photo by MSgt. Don Sutherland / National Archives*

The Thunderbirds team members are being filmed by local media prior to their air show performance in October 1987 at Osan Air Base, Republic of Korea. *National Archives*

During the 1987 Pacific tour, Lt. Col. Roger Riggs, the team leader, is conducting a planeside interview next to his F-16A aircraft with members of the Indonesia media. *USAF photo by MSgt. Don Sutherland / National Archives*

Crew chiefs stand by the cockpit as they give the one-minute warning signal prior to the start of the team's performance during the Thunderbirds Pacific tour in 1987. *USAF photo by MSgt. Don Sutherland / National Archives*

In July 1996, the Thunderbirds with their F-16C aircraft visited Lajes Field Air Base in the Azores, Portugal, as one of the stops that was made during the team's European tour of that year. *USAF photo by Senior Airman J. Arrowood / National Archives*

During the visit to the Azores in July 1996, the commander of the team, Lt. Col. Ron Mumm, answers questions about the Thunderbirds team with the local Portuguese media. *USAF photo by Senior Airman J. Arrowood / National Archives*

Thunderbird aircraft #5 has just completed aerial refueling over Germany en route to Timisoara, Romania, for an event there in June 1996. The team was touring former Eastern Bloc countries such as Romania, Bulgaria, and Slovenia. *USAF photo by TSgt. Brad Fallin / National Archives*

The Thunderbirds have just landed at Aviano Air Base, Italy, on July 3, 1996, in preparation for the air show held there. *USAF photo by TSgt. Hill-Wales / National Archives*

The Thunderbirds have just landed at Andersen AFB, Guam, in September 2004. This landing was the first time in a decade that the team had visited Guam. *USAF photo by SSgt. Bennie J. Davis III / National Archives*

Thunderbirds performing a line-abreast flat pass with smoke over Andersen AFB, Guam, in September 2004. *USAF photo by Airman 1st Class Kristin Ruleau / National Archives*

For the first time in fifty years, the US Air Force Thunderbird Demonstration Team performed in Colombia in July 2019, to over 250,000 Colombians during the Feria Aeronáutica Internacional at José María Córdova International Airport in Rionegro, Colombia, on July 14, 2019. *USAF photo by SSgt. Danny Randal*

The Thunderbirds perform a maneuver with the moon in the background during the air show held at José María Córdova International Airport in Rionegro, Colombia, on July 14, 2019. *USAF photo by SSgt. Danny Randal*

CHAPTER 14

Museum Displays and the Future

F-35A SPECIFICATIONS

Wingspan	35 feet
Length	50 feet, 6 inches
Height	14 feet, 3 inches
Empty weight	28,999 pounds
Power plant	one Pratt & Whitney F135 turbofan engine
Maximum speed	Mach 1.6
Service ceiling	50,000 feet
Range	1,700 miles

Since the F-16 continues to be a frontline fighter for the US Air Force, it works out for the Thunderbirds to continue to use the aircraft, as well as drawing Thunderbird pilots that trained on the F-16.

For the immediate future, the F-16C and D models used by the team will be receiving upgrades as part of a service life extension. After this time, the F-16 was originally scheduled to be phased out around 2025 in favor of the F-35. However, delays in the latter program have led to service life extension of the F-16 for another decade or more.

The potential replacement for the F-16 for the Thunderbirds could be the F-35A, which at this time is just starting to enter operational service. The F-35A is two generational steps above the F-16 in regard to the design. Pilot training on the F-35 is more intense than on the F-16, with knowledge being required to interface with the onboard computers and sensors.

In addition to seeing the Thunderbirds perform at different locations in the US and the world on a yearly basis, there are a number of displays of previous aircraft throughout the US as airport gate displays or in aircraft museums. Most notably, the USAF museums at Dayton and Warner Robbins are two that have vintage Thunderbird aircraft on display.

Regardless of the aircraft used by the team in the future, the mission of the team remains to plan and present precision aerial maneuvers to exhibit the capabilities of modern, high-performance aircraft and the high degree of professional skill required to operate those aircraft.

The F-35A is the new-generation fighter for the US Air Force that is reaching operational status. It is two generations in the design over the older F-16 aircraft, since the newer F-35A aircraft features many sensors and computers that are used by the pilots during flight. It remains to be seen if the Thunderbirds will eventually upgrade to this aircraft at some point in the future. *USAF*

The USAF Thunderbirds have their own dedicated museum, located at their home base of Las Vegas, Nevada. The museum features many photos of the team over the years, and posters and pilot gear used by the team in previous years. *John Gourley*

In addition to the wall displays and photos in the Thunderbirds museum, there are display cases containing used helmets signed by team members, as shown in the right photo, and on the left is a display case showing the different aircraft models that were flown by the team during its history, along with some of the support aircraft flown by the team. *John Gourley*

Members of the late F-105B Thunderbird pilot Capt. Gene Devlin's family, including his sons and wife, along with Rick Dale of Rick's Restoration, participate in a November 2013 ceremony dedicating a restored F-105B that is on public display in front of the Thunderbirds' hangar in Las Vegas, Nevada. *USAF photo by SSgt. Larry E. Reid Jr.*

This restored F-84F Thunderbird aircraft is on display at a public park located in Wachula, Florida, and has some of the markings of the Thunderbirds team still on the aircraft. There are Thunderbird aircraft displays that are located throughout the US at parks and airports. *Ken Neubeck*

Restored F-100D Thunderbird #6 aircraft, serial number 55-2754, is on display at the USAF museum located in Dayton, Ohio. *John Gourley*

Restored F-16A Thunderbird #1 aircraft, tail number 81-0663, is on display at the USAF museum located in Dayton, Ohio. *John Gourley*

The F-16 Thunderbirds have just completed a smoke emission flyover event with the British Red Arrows aero team, flying past the Statue of Liberty and over the Hudson River in New York City, headed north toward upstate New York, in August 2019. *Ken Neubeck*

A lone F-16 Thunderbird flown by Maj. Blaine Jones is performing a high alpha pass in July 2014 during the Arctic Thunder open house at joint base Elmendorf-Richardson, Alaska, with the mountains as a backdrop. The two-day show is performed in front of 20,000 people. *USAF photo by MSgt. John Nimmo Sr.*

On April 26, 2020, the USAF Thunderbirds teamed up with the US Navy Blue Angels for a special flyover of the New York metropolitan area for a show of support to the health care workers and first responders of New York during the Coronavirus pandemic that gripped the region. The two teams flew in twin delta formations, as seen in the top photo for much of the flight, and then side by side as seen in the bottom photo. The photos that were taken here were when the team flew over the eastern Long Island area on the way to Stony Brook University Hospital. The teams conducted aerial refueling at the start of the event.
Stu Walerstein